RELATIVE INVARIANTS OF SHEAVES

MONOGRAPHS AND TEXTBOOKS IN
PURE AND APPLIED MATHEMATICS

1. *K. Yano*, Integral Formulas in Riemannian Geometry (1970) *(out of print)*
2. *S. Kobayashi*, Hyperbolic Manifolds and Holomorphic Mappings (1970) *(out of print)*
3. *V. S. Vladimirov*, Equations of Mathematical Physics (A. Jeffrey, editor; A. Littlewood, translator) (1970) *(out of print)*
4. *B. N. Pshenichnyi*, Necessary Conditions for an Extremum (L. Neustadt, translation editor; K. Makowski, translator) (1971)
5. *L. Narici, E. Beckenstein, and G. Bachman*, Functional Analysis and Valuation Theory (1971)
6. *S. S. Passman*, Infinite Group Rings (1971)
7. *L. Dornhoff*, Group Representation Theory (in two parts). Part A: Ordinary Representation Theory. Part B: Modular Representation Theory (1971, 1972)
8. *W. Boothby and G. L. Weiss (eds.)*, Symmetric Spaces: Short Courses Presented at Washington University (1972)
9. *Y. Matsushima*, Differentiable Manifolds (E. T. Kobayashi, translator) (1972)
10. *L. E. Ward, Jr.*, Topology: An Outline for a First Course (1972) *(out of print)*
11. *A. Babakhanian*, Cohomological Methods in Group Theory (1972)
12. *R. Gilmer*, Multiplicative Ideal Theory (1972)
13. *J. Yeh*, Stochastic Processes and the Wiener Integral (1973) *(out of print)*
14. *J. Barros-Neto*, Introduction to the Theory of Distributions (1973) *(out of print)*
15. *R. Larsen*, Functional Analysis: An Introduction (1973) *(out of print)*
16. *K. Yano and S. Ishihara*, Tangent and Cotangent Bundles: Differential Geometry (1973) *(out of print)*
17. *C. Procesi*, Rings with Polynomial Identities (1973)
18. *R. Hermann*, Geometry, Physics, and Systems (1973)
19. *N. R. Wallach*, Harmonic Analysis on Homogeneous Spaces (1973) *(out of print)*
20. *J. Dieudonné*, Introduction to the Theory of Formal Groups (1973)
21. *I. Vaisman*, Cohomology and Differential Forms (1973)
22. *B. -Y. Chen*, Geometry of Submanifolds (1973)
23. *M. Marcus*, Finite Dimensional Multilinear Algebra (in two parts) (1973, 1975)
24. *R. Larsen*, Banach Algebras: An Introduction (1973)
25. *R. O. Kujala and A. L. Vitter (eds.)*, Value Distribution Theory: Part A; Part B: Deficit and Bezout Estimates by Wilhelm Stoll (1973)
26. *K. B. Stolarsky*, Algebraic Numbers and Diophantine Approximation (1974)
27. *A. R. Magid*, The Separable Galois Theory of Commutative Rings (1974)
28. *B. R. McDonald*, Finite Rings with Identity (1974)
29. *J. Satake*, Linear Algebra (S. Koh, T. A. Akiba, and S. Ihara, translators) (1975)

30. *J. S. Golan*, Localization of Noncommutative Rings (1975)
31. *G. Klambauer*, Mathematical Analysis (1975)
32. *M. K. Agoston*, Algebraic Topology: A First Course (1976)
33. *K. R. Goodearl*, Ring Theory: Nonsingular Rings and Modules (1976)
34. *L. E. Mansfield*, Linear Algebra with Geometric Applications: Selected Topics (1976)
35. *N. J. Pullman*, Matrix Theory and Its Applications (1976)
36. *B. R. McDonald*, Geometric Algebra Over Local Rings (1976)
37. *C. W. Groetsch*, Generalized Inverses of Linear Operators: Representation and Approximation (1977)
38. *J. E. Kuczkowski and J. L. Gersting*, Abstract Algebra: A First Look (1977)
39. *C. O. Christenson and W. L. Voxman*, Aspects of Topology (1977)
40. *M. Nagata*, Field Theory (1977)
41. *R. L. Long*, Algebraic Number Theory (1977)
42. *W. F. Pfeffer*, Integrals and Measures (1977)
43. *R. L. Wheeden and A. Zygmund*, Measure and Integral: An Introduction to Real Analysis (1977)
44. *J. H. Curtiss*, Introduction to Functions of a Complex Variable (1978)
45. *K. Hrbacek and T. Jech*, Introduction to Set Theory (1978)
46. *W. S. Massey*, Homology and Cohomology Theory (1978)
47. *M. Marcus*, Introduction to Modern Algebra (1978)
48. *E. C. Young*, Vector and Tensor Analysis (1978)
49. *S. B. Nadler, Jr.*, Hyperspaces of Sets (1978)
50. *S. K. Segal*, Topics in Group Rings (1978)
51. *A. C. M. van Rooij*, Non-Archimedean Functional Analysis (1978)
52. *L. Corwin and R. Szczarba*, Calculus in Vector Spaces (1979)
53. *C. Sadosky*, Interpolation of Operators and Singular Integrals: An Introduction to Harmonic Analysis (1979)
54. *J. Cronin*, Differential Equations: Introduction and Quantitative Theory (1980)
55. *C. W. Groetsch*, Elements of Applicable Functional Analysis (1980)
56. *I. Vaisman*, Foundations of Three-Dimensional Euclidean Geometry (1980)
57. *H. I. Freedman*, Deterministic Mathematical Models in Population Ecology (1980)
58. *S. B. Chae*, Lebesgue Integration (1980)
59. *C. S. Rees, S. M. Shah, and C. V. Stanojević*, Theory and Applications of Fourier Analysis (1981)
60. *L. Nachbin*, Introduction to Functional Analysis: Banach Spaces and Differential Calculus (R. M. Aron, translator) (1981)
61. *G. Orzech and M. Orzech*, Plane Algebraic Curves: An Introduction Via Valuations (1981)
62. *R. Johnsonbaugh and W. E. Pfaffenberger*, Foundations of Mathematical Analysis (1981)
63. *W. L. Voxman and R. H. Goetschel*, Advanced Calculus: An Introduction to Modern Analysis (1981)
64. *L. J. Corwin and R. H. Szcarba*, Multivariable Calculus (1982)
65. *V. I. Istrătescu*, Introduction to Linear Operator Theory (1981)
66. *R. D. Järvinen*, Finite and Infinite Dimensional Linear Spaces: A Comparative Study in Algebraic and Analytic Settings (1981)

67. *J. K. Beem and P. E. Ehrlich*, Global Lorentzian Geometry (1981)
68. *D. L. Armacost*, The Structure of Locally Compact Abelian Groups (1981)
69. *J. W. Brewer and M. K. Smith, eds.*, Emmy Noether: A Tribute to Her Life and Work (1981)
70. *K. H. Kim*, Boolean Matrix Theory and Applications (1982)
71. *T. W. Wieting*, The Mathematical Theory of Chromatic Plane Ornaments (1982)
72. *D. B. Gauld*, Differential Topology: An Introduction (1982)
73. *R. L. Faber*, Foundations of Euclidean and Non-Euclidean Geometry (1983)
74. *M. Carmeli*, Statistical Theory and Random Matrices (1983)
75. *J. H. Carruth, J. A. Hildebrant, and R. J. Koch*, The Theory of Topological Semigroups (1983)
76. *R. L. Faber*, Differential Geometry and Relativity Theory: An Introduction (1983)
77. *S. Barnett*, Polynomials and Linear Control Systems (1983)
78. *G. Karpilovsky*, Commutative Group Algebras (1983)
79. *F. Van Oystaeyen and A. Verschoren*, Relative Invariants of Rings: The Commutative Theory (1983)
80. *I. Vaisman*, A First Course in Differential Geometry (1984)
81. *G. W. Swan*, Applications of Optimal Control Theory in Biomedicine (1984)
82. *T. Petrie and J. D. Randall*, Transformation Groups on Manifolds (1984)
83. *K. Goebel and S. Reich*, Uniform Convexity, Hyperbolic Geometry, and Nonexpansive Mappings (1984)
84. *T. Albu and C. Năstăsescu*, Relative Finiteness in Module Theory (1984)
85. *K. Hrbacek and T. Jech*, Introduction to Set Theory, Second Edition, Revised and Expanded (1984)
86. *F. Van Oystaeyen and A. Verschoren*, Relative Invariants of Rings: The Noncommutative Theory (1984)
87. *B. R. McDonald*, Linear Algebra Over Commutative Rings (1984)
88. *M. Namba*, Geometry of Projective Algebraic Curves (1984)
89. *G. F. Webb*, Theory of Nonlinear Age-Dependent Population Dynamics (1985)
90. *M. R. Bremner, R. V. Moody, and J. Patera*, Tables of Dominant Weight Multiplicities for Representations of Simple Lie Algebras (1985)
91. *A. E. Fekete*, Real Linear Algebra (1985)
92. *S. B. Chae*, Holomorphy and Calculus in Normed Spaces (1985)
93. *A. J. Jerri*, Introduction to Integral Equations with Applications (1985)
94. *G. Karpilovsky*, Projective Representations of Finite Groups (1985)
95. *L. Narici and E. Beckenstein*, Topological Vector Spaces (1985)
96. *J. Weeks*, The Shape of Space: How to Visualize Surfaces and Three-Dimensional Manifolds (1985)
97. *P. R. Gribik and K. O. Kortanek*, Extremal Methods of Operations Research (1985)
98. *J.-A. Chao and W. A. Woyczynski, eds.*, Probability Theory and Harmonic Analysis (1986)
99. *G. D. Crown, M. H. Fenrick, and R. J. Valenza*, Abstract Algebra (1986)
100. *J. H. Carruth, J. A. Hildebrant, and R. J. Koch*, The Theory of Topological Semigroups, Volume 2 (1986)

101. *R. S. Doran and V. A. Belfi*, Characterizations of C*-Algebras: The Gelfand-Naimark Theorems (1986)
102. *M. W. Jeter*, Mathematical Programming: An Introduction to Optimization (1986)
103. *M. Altman*, A Unified Theory of Nonlinear Operator and Evolution Equations with Applications: A New Approach to Nonlinear Partial Differential Equations (1986)
104. *A. Verschoren*, Relative Invariants of Sheaves (1987)

Other Volumes in Preparation

RELATIVE INVARIANTS OF SHEAVES

A. Verschoren
University of Antwerp
Wilrijk, Belgium

MARCEL DEKKER, INC.   New York and Basel

ISBN 0-8247-7734-4

To my son Thomas (March 19, 1986),
who won't be able to read this
for a very, very long time...

PREFACE

There are several ways to introduce and to motivate the notion of relative invariants. One point of view is certainly the unifying aspect of the theory. Let me give an example to show what I mean by this. A rather natural invariant for commutative rings R is their so-called Picard group Pic(R) based upon isomorphism classes of invertible (= projective rank one) R–modules. This invariant is functorial, in the sense that for any ring map $f : R \to S$, there is an induced group map Pic(f) : Pic(R) $\to$ Pic(S), given by sending the isomorphism class [P] of an invertible R-module P to the isomorphism class $[P \otimes_R S]$ of $P \otimes_R S$, these maps satisfying the usual compatibility conditions with respect to the composition of morphisms. The multiplication in Pic(R) is induced by the tensor product $[P] \cdot [Q] = [P \otimes_R Q]$.

On the other hand, consider the class group Cl(R) of a noetherian Krull domain R, say. One of the possible alternatives is to define this group in terms of reflexive rank one R–modules, in a similar way as for the Picard group. Yet, in contrast with the case of invertible R–modules, the tensor product of a pair of reflexive rank one R–modules is not necessarily reflexive anymore. However, passing to the double dual does the trick; that is, if E and F are two reflexive rank one R–modules, then we take the modified tensor product $E \perp F$ to be $(E \otimes_R F)^{**}$, where $(-)^* = \mathrm{Hom}_R(- , R)$. One can show that $E \perp F = \cap (E \otimes F)_p$, where p runs through the height one prime ideals of R, the intersection being taken within $E \otimes F \otimes K$, where K denotes the field of fractions of R. In other words, if we denote by Q_1 the localization functor in the sense of Gabriel associated to the set of height one primes of R [denoted by $X^{(1)}(R)$], then $E \perp F = Q_1(E \otimes_R F)$. In particular, note that a torsion-free R–module E is reflexive if and only if $E = Q_1(E)$. One should also note that $E \perp E^* = R = Q_1(R)$, which shows that Cl(R) is a group indeed. The class group is not a functor, however, since in general a ring morphism $R \to S$ does not induce a group morphism $Cl(R) \to Cl(S)$.

It has been pointed out in [VV2] that other types of "class groups" or "generalized Picard groups" may also be defined (resp. recovered) in a similar way. Here one starts from a Gabriel topology $\mathbf{L}$ in R–mod, the category of R–modules, and (roughly speaking) one considers R–modules E that are (i) closed with respect to $\mathbf{L}$ [i.e., $E = Q_{\mathbf{L}}(E)$ if $Q_{\mathbf{L}}$ denotes the associated localization functor] and (ii) invertible with respect to $\mathbf{L}$ [i.e., there exists a similar R–module F such that $E \perp_{\mathbf{L}} F =: Q_{\mathbf{L}}(E \otimes_R F) = Q_{\mathbf{L}}(R)$]. The set of isomorphism classes of these forms a group $Pic(R, \mathbf{L})$ with product induced by $\perp$ and one easily verifies that most class group-type invariants that occur in practice are of this form. Most of the exact sequences, Mayer-Vietoris tricks, for example, which permit us to calculate and study the classical invariants occur as special cases of related exact sequences for these relative invariants.

The point of view adopted in [VV2] is that the study of the classical invariants, such as Pic, Cl, Br, benefits greatly from a uniform study of the relative invariants of which the former ones are special instances.

Another motivation for the sudy of relative invariants which was only briefly hinted at in [VV2] stems from algebraic geometry. Let us again start from an example. Consider a scheme X; then one may associate to it several invariants, such as its Picard group $Pic(X)$ or its Brauer group $Br(X)$. However, these invariants are not essentially new when X is affine, $X = Spec(R)$ say, for in this case $Pic(X) = Pic(R)$ [resp., $Br(X) = Br(R)$, etc.], so eveything reduces to the ring and module case.

On the other hand, even if we start from an affine scheme $Spec(R)$ and we take an open subscheme $U \subset Spec(R)$, the calculation of $Pic(U)$, for example, is not trivial anymore, essentially because quasicoherent sheaves on U are not in one-to-one correspondence with modules over the coordinate ring. However, since one understands pretty well what happens on $Spec(R)$, it would be useful if one were able to (i) extend invertible sheaves on U to "nice" (at least quasicoherent) sheaves on Spec(R), and (ii) classify these possible extensions, for in this case the determination of Pic(U) would reduce to a calculation of a "relative invariant" of Spec(R) (with respect to U), based upon certain well-determined R–modules. This problem is closely related to some results of Horrocks' [Hor], who studied the classification of bundles (= locally free coherent sheaves) on the punctured spectrum of a local ring R and their possible extensions to the whole scheme.

Of course, in practice, the problems one encounters are in general more complicated. Indeed, one usually starts from a "nice" scheme X (e.g., a locally noetherian normal scheme) and one looks at an arbitrary, not necessarily open, subset Y of X, endowed with the induced sheaf $\underline{O}_Y = \underline{O}_X|Y$. One then wants to calculate invariants for the ringed space $(Y, \underline{O}_Y)$. This is necessary, for example, if one wants to calculate the divisor class group of a normal scheme, for then one should know $\text{Pic}(X^{(1)}, \underline{O}_X|X^{(1)})$, where $X^{(1)}$ is the subset of codimension one points of X. In this case one is of course also primarily interested in possible extensions of (quasicoherent) sheaves on Y to (quasicoherent) sheaves on X. In contrast with the case of open subschemes, extension results of this more general type are not widely available in the literature. One of the purposes of this text is to fill in the gap. It appears moreover that to make our results work for Krull schemes for instance, one has to weaken considerably the noetherian hypotheses which are subsumed even in the classical situation. The problem we thus have briefly sketched (i.e., that of extending sheaves and classifying their extensions) will be dealt with in Chapter 6. As it happens, the subsets of a scheme X that we are really interested in are the so-called generically closed subsets (we refer to the text for a proper definition). In the affine case, that is, X = Spec(R) for some ring R (which we assume to be noetherian for the moment), these generically closed subsets are in bijective correspondence with Gabriel topologies in R–mod. Since it is therefore clear that these Gabriel topologies will play a leading role in our setup, we have included in Chapter 5 a rather self-contained introduction to the subject. Of course, the topics covered in that chapter were chosen from a strictly utilitarian point of view : only results used and needed in the sequel are mentioned. We have tried to eliminate noetherian assumptions whenever possible and besides several recent results, we have included new proofs for older results dispersed in the literature. Although no acquaintance with the contents of [VV2], say, is strictly needed for an understanding of the present text, one may view some of the results in Chapter 5 as a generalization as well as a strengthening of similar results covered (under stronger hypotheses) in that source.

The results in Chapter 5 are then used in Chapter 6 to derive the announced extension properties for coherent and quasicoherent sheaves. In the affine situation some partial results (in the noetherian case) were already briefly sketched in [VV2], for example. Here we deal with the general case, that is, we do not assume our schemes to be (locally) noetherian or affine. In an appendix we also provide some connections between our theory and local cohomology (as developed by Suominen [Su], for example). Of course, in order to follow the development so far, the reader is assumed to possess a reasonable background in sheaves and elementary algebraic geometry. I have tried to make the text as self-contained as possible (within certain limits) by including some introductory chapters on general sheaf theory, which may be viewed as a "shortest way to the extension theorems." These chapters may be skipped by readers sufficiently acquainted with the fundamentals of sheaf theory, or, alternatively, briefly glanced through, to refresh one's memory. These chapters may also be used independently as an introductory course on general sheaf theory (as I did sometimes) in combination with other, more advanced material. However, let me point out that I have avoided treating the general (Grothendieck) cohomology of sheaves here (mainly because it is not needed or used in the sequel), sticking to down-to-earth Cech cohomology instead, whenever this was possible.

Global relative invariants are treated in Chapter 7, together with the fundamental notion of torsion couples and morphisms between them, these being used afterward to derive and explain the functorial features of relative and classical invariants with respect to special types of morphisms. The invariants introduced here are connected through exact sequences, generalizing similar sequences due to Auslander [Au1], Lee-Orzech [LO1], Orzech [Or1], Van Oystaeyen-Verschoren [VV2], Yuan [Yu], et al. Answering a question raised in [LO1], we prove in an appendix how these sequences may also be derived by K-theoretic means.

This also prepares for Chapter 8, where we specialize to Krull schemes and sheaves over them. For obvious reasons, divisorial sheaves play a central role here, so we start off by linking these notions to some torsion theoretic concepts. In this way some results due to [Ha2, OSS] and relating to normal sheaves may be generalized and strengthened. We also provide links with local cohomology along with some indications of the relationship between our results and some results due to M. Orzech [Or1], which say that an extension of Krull domains S/R has PDE if and only if S is a divisorial R-module.

The final chapter is devoted to Hecke actions. Here we study contravariant maps between relative invariants, which generalize the norm maps introduced by Knus and Ojanguren in the classical case. Although one may construct these norm maps by hand by a straightforward descent argument (as we do), it appears that the Hecke actions introduced by Roggenkamp and Scott [RS] are just the right tool to study these. The results obtained in this chapter are closely related to similar ones in algebraic number theory, which are used there to calculate class groups of number fields or more generally of arbitrary Dedekind domains. We have preferred to restrict our discussion essentially to finite extensions, that is, all Galois extensions occurring in our setup are finite. Some indications about the infinite case (with references to the literature) are also provided to the interested reader, however.

Note : This text is not Riri 3 (where Riri is the working title F. Van Oystaeyen and I use for the Relative Invariants of Rings series) and may be read completely independently of it - although some acquaintance with the previous Riri volumes is certainly helpful when reading through the present volume.

Finally, it is my pleasure to acknowledge support by the Nationaal Fonds voor Wetenschappelijk Onderzoek (N.F.W.O.), while I was working on this text. I also thank the University of Antwerp (Universitaire Instelling Antwerpen and Rijksuniversitair Centrum Antwerpen) for its continuing hospitality, as well as the Universities of Granada, Valencia, and Murcia (where I gave some talks on the contents of this book) for their short-term hospitality. I acknowledge the criticism of my Spanish friends and colleagues, J. L. Bueso, B. Torrecillas, P. Jara, J. L. Gomez Pardo, D. Tarrazona, Maribel Segura, and "el Tigre" during these talks, who thereby helped in shaping the final version of this text. I also wish to thank Freddy Van Oystaeyen (for friendship and support), A. Beckelheimer (for some critical remarks), and last but not least, my students, on whom I tried out the first chapters of this book (without their knowing it), and my colleagues at U. I. A., who (in their own way) stimulated me to write this text.

This book was written on an Apple Macintosh computer, using Microsoft Word. If it weren't for Dirk Fabré (who helped me a lot), its final shape would have been completely different.

A. Verschoren

CONTENTS

RELATIVE INVARIANTS OF SHEAVES

1 GENERALITIES ON SHEAVES

(1.1.) The main purpose of this Section is to present in a more or less self-contained way some of the necessary sheaf-theoretic machinery and terminology, which will be used throughout in the sequel. For more and detailed information, we refer the reader to the many excellent textbooks on this material, such as Bredon [Br], Godement [Go], Swan [Sw1], Tennisson [Te], ... Specific information concerning structure sheaves on affine and other schemes will be given and dealt with in the next Sections.

(1.2.) The easiest and perhaps the most natural way to introduce sheaves is through the notion of sheaf spaces (or "espaces étalés" in the terminology of Godement [Go] et al.). Let X and E be topological spaces and let $p : E \to X$ denote a continuous map. The triple $\mathbf{E} = (E, X, p)$ is called a <u>sheaf space</u> if p is a local homeomorphism. This means that for any $e \in E$ we may find an open neighborhood U of e in E and an open neighborhood V of $p(e)$, such that the induced map $p|U : U \to V$ is a homeomorphism. We call X the <u>base space</u> of $\mathbf{E}$ and p the <u>projection</u> of $\mathbf{E}$. If no ambiguity arises, we will simply say that E is a sheaf space.

(1.3.) Example Let E denote the helix in $\mathbf{R}^3$ parametrized by

$$\mathbf{R} \to E \ : \ t \to (t, \sin(t), \cos(t))$$

and with topology induced by that of $\mathbf{R}^3$, then the projection morphism

$$p : E \to \mathbf{R}^2 \ : \ (t, \sin(t), \cos(t)) \to (\sin(t), \cos(t))$$

induces a sheaf space (E, X, p), where $X = \mathbf{S}^1$, the unit circle in $\mathbf{R}^2$, also with the induced topology. Indeed, the restriction of p to X and E is obviously a local homeomorphism (but not a homeomorphism globally!).

(1.4.) Let A be an arbitrary set, then for any topological space X we may consider the product space $A \times X$, where A is given the discrete topology. The triple $(A \times X, X, p)$, where $p : A \times X \to X$ is the projection onto the second factor, is then obviously a sheaf space. We denote this by A_X and we call it the <u>constant sheaf space</u> with fibre A on X.

(1.5.) If $\mathbf{E} = (E, X, p)$ is a sheaf space, the we denote by $E_x = p^{-1}(x)$ the <u>stalk</u> of $\mathbf{E}$ (or E) over or at x. It is clear that E and the disjoint union $\cup\{E_x ; x \in X\}$ may be identified as sets. We define a section of $\mathbf{E}$ (or E) over an open subset U of X to be a continuous map $s : U \to E$ such that $ps = id_U$. The collection of all sections of E over a fixed open subset U of X is denoted by $\Gamma(U, \mathbf{E})$ (or even by $\Gamma(U, E)$). An element $s \in \Gamma(X, E)$ is called a <u>global section</u> of $\mathbf{E}$ (or E).

(1.6.) Exercise 1. Let (E, X, p) be the sheaf space described in (1.3.). Show that for any non empty open subset U of X, there is a natural identification $\Gamma(U, E) = \mathbf{Z}$. Show also that $E_x = \mathbf{Z}$ for every $x \in X$.
2. Let A_X be the constant sheaf space with fibre A on X. Show that for any non empty open subset U of X, there is a natural identification $\Gamma(U, A_X) = A^{c(U)}$, where c(U) denotes the set of connected components of U.

(1.7.) Let X be a Hausdorff space, let $x_0 \in X$ and define

$$E = (X - \{x_0\}) \times \{0\} \cup \{x_0\} \times \mathbf{Z} \subset X \times \mathbf{Z}.$$

Let B consist of the subsets of E of the form $U \times \{0\}$, with U open in X and $x_0 \notin U$ or of the form $(V - \{x_0\}) \times \{0\} \cup \{(x_0, z)\}$, where V is open in X and x_0

$\in$ V and z $\in$ **Z**. Clearly, B is the basis for a topology on X, and together with the canonical projection, this yields the so-called <u>skyscraper sheaf space</u> (E, X, p) on X with support x_0 and fibre **Z**. If x $=x_0$ in X, then $E_x =$ **Z** and otherwise $E_x = \{0\}$. Moreover, if U is open in in X, then it is easy to see that $\Gamma(U, E) =$ **Z** if $x_0 \in$ U, while $\Gamma(U, E) = \{0\}$ in the other case.

The proof of the following result is left as an exercise to the reader:

(1.8.) Proposition Let (E, X, p) be a sheaf space, then

(1.8.1.) the map p : E $\rightarrow$ X is open;

(1.8.2.) if U $\subset$ X is open and s $\in \Gamma(U, E)$, then s(U) is open in E;

(1.8.3.) the collection of all s(U), where U $\subset$ X is open and s $\in \Gamma(U, E)$, forms a basis for the topology of E.

(1.9.) Corollary If (E, X, p) is a sheaf space and U, V are open subsets of X, then for any s $\in \Gamma(U, E)$ and t $\in \Gamma(V, E)$ the set of all x $\in$ U $\cap$ V such that s(x) = t(x) is open in X.

Proof Indeed, the foregoing Proposition shows that s(U) $\cap$ t(V) is open in E, so s and t agree on p(s(U) $\cap$ t(V)), which is open by (1.8.1.).

(1.10.) It follows from this that a sheaf space is in general not necessarily a Hausdorff space. Indeed, let (E, X, p) be a sheaf space, then for any open subset U $\subset$ X and any pair of sections s, t of E over U, it follows that the set W, consisting of all x in U such that s(x) = t(x), is open in U. But, if E is a Hausdorff space, then W is also closed in U. This implies that if s and t agree in one point of U, then they agree on the connected component of this point, i.e. E satisfies the so-called principle of "analytic continuation". However, from the fact that sheaf spaces and sheaves may be identified, as we will see below, it follows that this principle is far from being valid in general (take the sheaf of germs of continuous functions on **C**, for example).

(1.11.) Let X be a topological space and $x \in X$. We will write $V(x)$ or $V_X(x)$ for the open neighborhood filter of x, ordered by putting $U < V$ if $V \subset U$. If **E** = (E, X, p) is a sheaf space, then the family

$$\{\; \Gamma(U, E),\; \varepsilon^U_V;\; U, V \in V(x),\; U < V\}$$

is an inductive system, for any $V \subset U$ the map $\varepsilon^U_V : \Gamma(U, E) \to \Gamma(V, E)$ being defined by sending a section $s : U \to E$ tot its restriction $s|V : V \to E$. Denote by **E**(x) or E(x) the inductive limit

$$\lim_{U \in V(x)} \Gamma(U, E) = \lim \Gamma(\,\text{-}\,, E).$$

(1.12.) Proposition For any sheaf space (E, X, p) and any $x \in X$, there is a canonical bijection $E_x \cong E(x)$.

Proof Let $U \in V(x)$, then evaluating sections yields a map $\varepsilon^U_x : \Gamma(U, E) \to E_x : s \to s(x)$. If $U < V$, i.e. $V \subset U$, then there is a commutative diagram

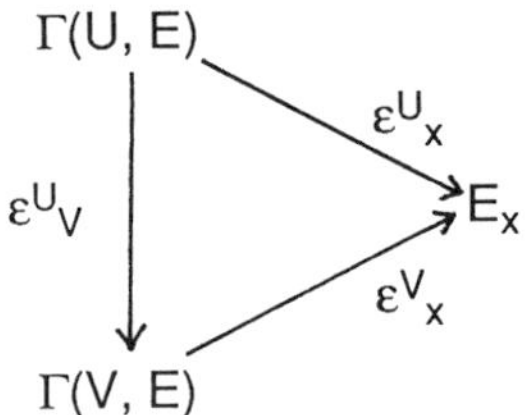

and it follows that there is a map $\theta : E(x) = \lim \Gamma(U, E) \to E_x$.

Write $\varepsilon^U : \Gamma(U, E) \to \lim \Gamma(U, E) = E(x)$ for the canonical map, then for any open neighborhood U of x, the map θ makes the following diagram commutative:

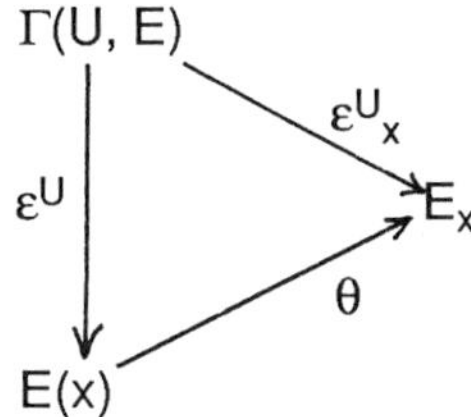

We claim that θ is bijective. Indeed, to show that θ is injective, take s and t in E(x) with $\theta(s) = \theta(t)$, then by definition there exist U and V in V(x) and some $s \in \Gamma(U, E)$ resp. $t \in \Gamma(V, E)$ with the property that $\sigma = \varepsilon^U(s)$ resp. $\tau = \varepsilon^V(t)$. Since $\theta(\sigma) = \theta(\tau)$, it follows from the commutativity of the above diagram that $\varepsilon^U_x(s) = \varepsilon^V_x(t)$, i.e. $s(x) = t(x) \in E_x$. From (1.9.) it now follows that there exists some $W \subset U \cap V$ in V(x), such that $\varepsilon^U_W(s) = s|W = t|W = \varepsilon^V_W(t)$, hence $\sigma = \varepsilon^U(s) = \varepsilon^W \varepsilon^U_W(s) = \varepsilon^W \varepsilon^V_W(t) = \varepsilon^V(t) = \tau$ and θ is injective, indeed! On the other hand, to show that θ is surjective, pick $e \in E_x$, then e possesses an open neighborhood U in E such that for some open subset V of X, the map $p|U : U \to V$ is a homeomorphism. The inverse map $s = (p|U)^{-1} : V \to U \subset E$ has the property that $s \in \Gamma(V, E)$ and $s(x) = e$. Put $\sigma = \varepsilon^V(s)$, then we find that $\theta(\sigma) = \theta\varepsilon^V(s) = s(x) = e$ and from this we derive that θ is surjective.

(1.13.) Let us now define the notion of a morphism of sheaf spaces, in order to make the collection of these into a category. So, assume that $\mathbf{E} = (E, X, p)$ and $\mathbf{E'} = (E', X, p')$ are sheaf spaces, then a <u>morphism</u> $f : \mathbf{E} \to \mathbf{E'}$ is by definition a map $f : E \to E'$, which makes the following diagram commutative:

(1.13.1.)

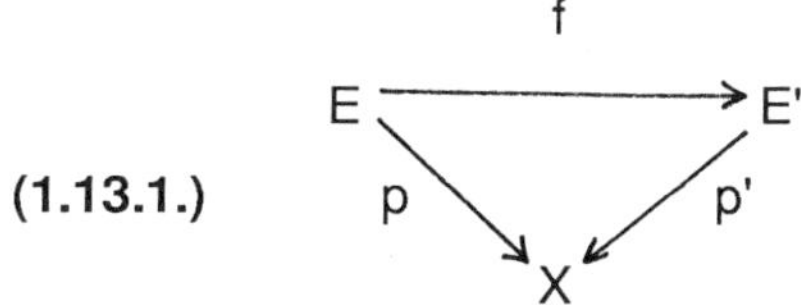

and which is continuous.

Exercise Prove, in the presence of the above commutativity condition, that this is equivalent to f being open or f being a local homeomorphism.

Morphisms of sheaf spaces are composed in the obvious way and this yields a category $\mathbf{Sh}(X)$, whose objects are the sheaf spaces with base space X and whose morphisms are the morphisms of sheaf spaces we just defined.

It is clear that if $f : \mathbf{E} \to \mathbf{E'}$ is a morphism of sheaf spaces, then for any x in X, this morphism induces a map $f_x : \mathbf{E}_x \to \mathbf{E'}_x$ between stalks (due to the commutativity of the diagram (1.13.1.) above) and this is easily seen to define a covariant functor $\tau_x : \mathbf{Sh}(X) \to \mathbf{Sets}$, sending $\mathbf{E}$ tot its stalk $\mathbf{E}_x$ in x and any $f : \mathbf{E} \to \mathbf{E'}$ to its induced map $f_x : \mathbf{E}_x \to \mathbf{E'}_x$.

Similarly, for any open subset U of X, a morphism of sheaf spaces $f : \mathbf{E} \to \mathbf{E'}$ yields a map $\Gamma(U, f) : \Gamma(U, \mathbf{E}) \to \Gamma(U, \mathbf{E'}) : s \to fs$, and again, since it is clear that $\Gamma(U, gf) = \Gamma(U, g)\Gamma(U, f)$ for any pair of morphisms $f : \mathbf{E} \to \mathbf{E'}$ resp. $g : \mathbf{E'} \to \mathbf{E''}$, it follows that this yields a covariant functor $\Gamma(U, -) : \mathbf{Sh}(X) \to \mathbf{Sets}$ for any open subset U of X.

(1.14.) Example (sheaf subspaces) Let (E, X, p) be a sheaf space and F an open subspace of E; let q = p|F, then we claim that (F, X, q) is a sheaf space as well. Indeed, it suffices to check that q is a local homeomorphism, so, let $e \in F$, then we may find an open neighborhood V of e in E and an open neighborhood U of $p(e) = q(e) \in X$ such that p|V : $V \to U$ is a homeomorphism too, and $V \cap F$ is open in F resp. $p(V \cap F)$ open in X. This proves the assertion. Moreover, from the Exercise in (1.13) it follows that the inclusion $i : F \subset E$ induces a morphism of sheaf spaces $i : (F, X, q) \to (E, X, p)$. We say that F is a <u>sheaf subspace</u> of E.

The functor $\Gamma(U, -)$ yields for any open subset U of X a map $\Gamma(U, i)$: $\Gamma(U, F) \to \Gamma(U, E)$, which is obviously injective (since i is injective). Conversely, if a morphism of sheaf spaces $i : (F, X, q) \to (E, X, p)$ induces for any open subset U of X an injective map $\Gamma(U, F) \to \Gamma(U, E)$, then we claim that i is injective, hence that F may be identified with an open subset of E. Indeed, let $f, f' \in F$ and assume $i(f) = i(f')$, then $q(f) = q(f') = x$, say, because $pi = q$. We may pick an open neighborhood U of x in X and $s, s' \in \Gamma(U, F)$ such that $s(x) = f$ resp. $s'(x) = f'$. But then $\Gamma(U, i)(s)(x) = is(x)$ $= i(f) = i(f') = is'(x) = \Gamma(U, i)(s')(x)$, i.e. the sections $\Gamma(U, i)(s)$, $\Gamma(U, i)(s') \in$ $\Gamma(U, E)$ coincide in $x \in X$. It follows that we may find $V \in V(x)$ contained in U such that $\Gamma(U, i)(s)|V = \Gamma(U, i)(s')|V$. But $\Gamma(U, i)(s)|V = \Gamma(V, i)(s|V)$ and similarly for $\Gamma(U, i)(s')|V$, hence the injectivity of $\Gamma(V, i)$ shows that $s|V =$ $s'|V$, so in particular $f = s(x) = s'(x) = f'$, i.e. i is injective, indeed.

(1.15.) Denote by **Open**(X) the category defined as follows: the objects of **Open**(X) are the open subsets of X (= Open(X)), the morphisms the canonical inclusions $V \subset U$. For any sheaf space $\mathbf{E} = (E, X, p)$ there is a contravariant functor

$$\Gamma(- , \mathbf{E}) = \Gamma(- , E) : \mathbf{Open}(X) \to \mathbf{Sets},$$

which associates to any U in **Open**(X) the set of sections $\Gamma(U, E)$ and with any inclusion $V \subset U$ the morphism

$$\varepsilon^U_V = \mathrm{res}^U_V : \Gamma(U, E) \to \Gamma(V, E) : s \to s|V$$

This is a contravariant functor indeed, since for any U in **Open**(X) we have $\mathrm{res}^U_U = \mathrm{id}_{\Gamma(U, E)}$ and since for any open sets $W \subset V \subset U$ we have $\mathrm{res}^U_W = \mathrm{res}^V_W \mathrm{res}^U_V$.

The functor $\Gamma(-, E)$ is an example of a presheaf on X. As we will see below, one of the first ideas in sheaf theory, will be to realize any presheaf geometrically, i.e. to associate in a natural way a sheaf space to it, with "enough" "nice" properties.

(1.16.) Let us start with a definition. Formally, a <u>presheaf</u> (of sets) for a topological space X is determined by the following data:

(1.16.1.) a basis B for the topology on X;

(1.16.2.) a collection of sets $\{E(U); U \in B\}$;

(1.16.3.) a collection of maps $E^U_V : E(U) \to E(V)$, one for each pair of open subsets $V \subset U$ in B.

These data have to satisfy the following properties:

(*) $E^U_U = id_{E(U)}$ for each $U \in B$;

(**) if $W \subset V \subset U$ in B, then $E^V_W E^U_V = E^U_W$:

The collection B may be viewed as a full subcategory **B** of **Open**(X) and then the above definition just says that a presheaf (of sets) is a contravariant functor

$$E : \mathbf{B} \to \mathbf{Sets},$$

which to any $U \in B$ associates the set $E(U)$ and to any $V \subset U$ in B the <u>restriction</u> $E^U_V : E(U) \to E(V)$. We usually refer to E as a <u>presheaf on</u> B. If $B = Open(X)$, then we will say that E is a presheaf on X.

(1.17.) Example (the presheaf associated to a sheaf space). Let $\mathbf{E} = (E, X, p)$ be a sheaf space, then for any basis B for the topology on X, the data

$$\{\Gamma(U, E), \varepsilon^U_V; U, V \in B, V \subset U\}$$

determine a presheaf on B. We denote this presheaf by $\Gamma(-, E)$ and call it the <u>presheaf of sections</u> of **E** or E (on B).

(1.18.) Example (the constant presheaf). Let B a basis for the topology on X and A an arbitrary set, then we may define a presheaf $c_B(A)$ on B by putting $c_B(A)(U) = A$ for any $U \in B$ and $c_B(A)^U_V = id_A$ for any $V \subset U$ in B. We call this the <u>constant presheaf</u> with fibre A on B.

(1.19.) Example Let X and Y be topological spaces and B a basis for the topology on X. For any $U \in B$ let $C^Y(U)$ consist of all continuous maps $f : U \to Y$ and for any $V \subset U$ in B, put $(C^Y)^U_V : C^Y(U) \to C^Y(V) : f \to f|V$. This defines a presheaf, which we call the <u>presheaf (of germs) of Y-valued continuous functions</u> on B. If we put more structure on X, say X is an algebraic or differentiable manifold, and if we choose corresponding functions f, say f is regular or C^∞, with $Y = \mathbf{C}$, then we obtain the presheaves of regular, differentiable, ... functions on B.

(1.20.) Example (the structure presheaf on an affine scheme). We will frequently have to deal with the following example. The set of all prime ideals Spec(R) of a commutative ring R may be endowed with the so-called Zariski topology, i.e. with open sets of the form $X(S) = \{p \in \text{Spec}(R); S \subset p\}$ for some subset S of R, which may be chosen to be an ideal of R. Since for any pair of ideals I, J resp. any collection of ideals $\{I_\alpha; \alpha \in A\}$ of R we clearly have $X(I) \cap X(J) = X(IJ)$ resp. $\cup_\alpha X(I_\alpha) = X(\Sigma_\alpha I_\alpha)$, it follows that this is a toplogy indeed. It is also clear that for any pair of ideals I, J of R we have $X(I) = X(rad(I))$ and $X(I) \subset X(J)$ if and only if $rad(I) \subset rad(J)$, where rad(I) is the radical of I, i.e. the intersection of all prime ideals of R containing I or equivalently the ideal of all r in R with $r^n \in I$ for some positive integer n. The sets X(s) with $s \neq 0$ in R form a basis B for the Zariski topology on Spec(R). If $X(t) \subset X(s)$, then $t \in rad(Rs)$, i.e. $t^n = rs$ for some positive integer n and some $r \in R$. We define a presheaf E on B by putting $E(X(s)) = R_s$, the ring of fractions of R at the multiplicative subset of R generated by s and by defining for any $X(t) \subset X(s)$, i.e. with t^n = rs as above, the restriction $E(X(s)) \rightarrow E(X(t))$ to be the map $R_s \rightarrow R_t$ which sends $u/s^m \in R_s$ tot $ur^m/r^m s^m = ur^m/t^{nm} \in R_t$. The presheaf E will be denoted by $\underline{Q}_R$ and will be called the <u>structure presheaf</u> on B.

(1.21.) Proposition (1.12.) above yields a natural way to define "stalks" for an arbitrary presheaf. Indeed, let E be a presheaf on a basis B of the topological space X and put $V_B(x) = V(x) \cap B$, the filter of open neighborhoods of x in B. The set $\{E(U), E^U_V; U, V \in V_B(x), V \subset U\}$ forms an inductive system. We write

$$E(x) = \lim_{U \in V_B(x)} E(U),$$

and we call this the <u>stalk</u> of E in x.

In analogy with the terminology used for sheaf spaces, we sometimes call the elements of E(U) <u>sections</u> of E over U. Therefore, the elements of E(x) are usually referred to as <u>germs</u>(of sections) of E.

(1.22.) Examples Let us take a look at the stalks of the presheaves described in the examples above.

(1.22.1.) Let $\Gamma(- , E)$ denote the presheaf of sections of the sheaf space (E, X, p), then the stalk $\Gamma(- , E)(x)$ of $\Gamma(- , E)$ at x may be identified (by (1.12.)!) with the stalk E_x of E at x.

(1.22.2.) If $c_B(A)$ is the constant presheaf with fibre A on B, then for any $x \in X$, the stalk $c_B(A)(x)$ may be identified with the set A.

(1.22.3.) Let C^Y denote the presheaf of (germs of) Y-valued continuous functions on some basis B of the topological space X. For any $x \in X$, the stalk $C^Y(x)$ consists exactly of the germs in x of Y-valued continuous functions, i.e. it consists of the equivalence classes of pairs (U, f), where $U \in V_B(x)$ and $f : U \to Y$ is continuous, any two such couples (U, f), (V, g) being equivalent if and only if for some $W \subset U \cap V$ in $V_B(x)$ we have $f|W = g|W$.

(1.22.4.) Let R be a commutative ring and let B be the basis for the Zariski topology on $\mathrm{Spec}(R)$ described in (1.20.). If $p \in \mathrm{Spec}(R)$, then we claim that the stalk in p of the structure presheaf $\underline{Q}_R$ on B is given by $\underline{Q}_R(p) = R_p$, the ring of fractions of R at the prime ideal p. Indeed, for any $X(s) \in B$ containing p, by definition $s \notin p$, hence there is a canonical map $\rho_s : \underline{Q}_R(X(s)) = R_s \to R_p$. These maps being compatible for different s, they yield a map $\rho : \underline{Q}_R(p) = \lim \underline{Q}_R(X(s)) \to R_p$. We want to show that ρ is bijective. Now, ρ is clearly surjective, since for any $\alpha = u/s \in R_p$, i.e. with $u \in R$ and $s \in R - p$, we have $\alpha = \rho_s(u/s)$, where u/s may now be considered as an element of R_s. On the other hand, take $\beta \in \underline{Q}_R(p)$ represented by some $u/s \in R_s$, and assume that $\rho(\beta) = 0$, then $\rho_s(u/s) = 0$, i.e. $u/s = 0$ in R_p. It follows that we may find $t \in R - p$ with $tu = 0$. But then, $tu/ts = 0$ in R_{ts} and tu/ts also represents $\beta \in \underline{Q}_R(p)$, so $\beta = 0$, indeed. Since ρ is obviously a ring morphism, this proves that ρ is also injective.

(1.23.) Let E be a presheaf on B, let $U \in B$ and $x \in X$, then we denote by $E^U_x : E(U) \to E(x) = \lim E(U)$ the canonical morphism. For any germ e in $E(x)$, there exists an $s \in E(U)$ for some $U \in V_B(x)$, for which $e = E^U_x(s)$ (due to the definition of inductive limits!). We will denote this element $E^U_x(s) \in E(x)$ by s_x. It is thus clear that for any $U, V \in V_B(x)$ and any $s \in E(U)$ resp. $t \in E(V)$, we have $s_x = t_x \in E(x)$ if and only if we may find some $W \in V_B(x)$ with $W \subset U \cap V$ and $E^U_W(s) = E^V_W(t)$.

Note also that the stalks of a presheaf are "independent" of the basis of open sets on which it is defined. Indeed, if $B' \subset B$ are both a basis of open sets for the topology on X and if $x \in X$, then for any presheaf E on B we have that

$$\lim_{U \in V_{B'}(x)} E(U) \quad = \quad \lim_{U \in V_B(x)} E(U)$$

This follows immediately from the fact that $V_{B'}(x)$ is cofinal in $V_B(x)$. In particular, if E is a presheaf on X and B a basis of open sets of X, then

$$\lim_{U \in V_B(x)} E(U) \quad = \quad \lim_{U \in V_X(x)} E(U)$$

(1.24.) Let us now make the collection of presheaves on a fixed basis B of a topological space X into a category **P**(B). If E and F are presheaves on B, then a <u>morphism</u> g : E → F is defined by giving for each U in B a map g(U) : E(U) → F(U) with the property that for each V ⊂ U in B, the following diagram commutes:

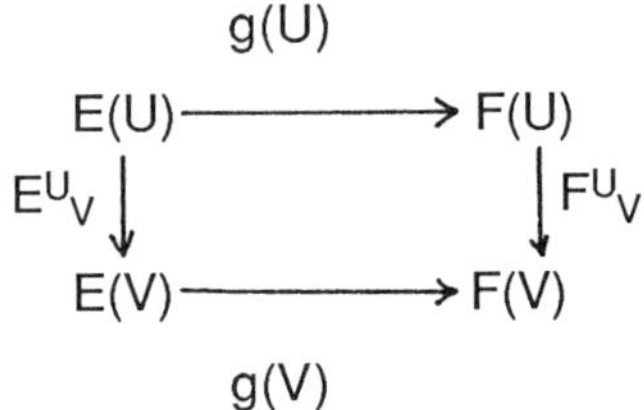

In other words, a morphism of presheaves is just a natural transformation between the underlying functors. If g : E → F and h : F → G are morphisms between presheaves on B, then the composition hg : E → G is defined by putting (hg)(U) = h(U)g(U) for every U in B. This obviously yields a category **P**(B), the <u>category of presheaves</u> (of sets) on B. We write **P**(X) for **P**(Open(X)), the category of presheaves (of sets) on X. Morphisms in **P**(B) behave very nicely, e.g. one easily shows that a morphism f in **P**(B) is an isomorphism if and only if f(U) is an isomorphism for any U in B.

(1.25.) Let $\mathbf{E} = (E, X, p)$ and $\mathbf{F} = (F, X, q)$ be sheaf spaces. If $f : E \to F$ is a morphism of sheaf spaces, then f induces for any open subset U of X a map $\Gamma(U, f) : \Gamma(U, E) \to \Gamma(U, F) : s \to fs$, such that for any open subsets $V \subset U$ the following diagram commutes:

$$\begin{array}{ccc}
 & \Gamma(U, f) & \\
\Gamma(U, E) & \longrightarrow & \Gamma(U, F) \\
E^U{}_V \downarrow & & \downarrow F^U{}_V \\
\Gamma(V, E) & \longrightarrow & \Gamma(V, F) \\
 & \Gamma(V, f) &
\end{array}$$

Here $E^U{}_V$ and $F^U{}_V$ are the restriction morphisms for E resp. F. It follows that every f in $\mathrm{Hom}_{\mathbf{Sh}(X)}(E, F)$ yields a morphism

$$\Gamma(\,-\,, f) \in \mathrm{Hom}_{\mathbf{P}(B)}(\Gamma(\,-\,, E), \Gamma(\,-\,, F))$$

for every basis B of open subsets of X. It follows:

(1.26.) Proposition Let X be a topological space and B a basis of open sets of X, then there is a canonical functor $\Gamma : \mathbf{Sh}(X) \to \mathbf{P}(B)$, mapping any sheaf space $\mathbf{E}$ to the presheaf $\Gamma(\,-\,, \mathbf{E})$ on B and any morphism of sheaf spaces $f : \mathbf{E} \to \mathbf{F}$ to $\Gamma(\,-\,, f) : \Gamma(\,-\,, \mathbf{E}) \to \Gamma(\,-\,, \mathbf{F})$.

(1.27.) We will now be concerned with the following question : which presheaves are of the form $\Gamma(\,-\,, \mathbf{E})$ for some sheaf space $\mathbf{E}$ on X, i.e. we want to describe the image of Γ in $\mathbf{P}(B)$. It appears that this problem is closely related to that of constructing a left adjoint to Γ. One of the main ingredients we will need is that for any presheaf E on a basis B of the topology on X and for any $x \in X$, there is a canonical functor $\mathbf{P}(B) \to \mathbf{Sets}$, which we will also denote by τ_x. This functor associates to any E in $\mathbf{P}(B)$ the stalk $E(x) = \lim E(U)$, where U runs through $V_B(x)$ and to any morphism of presheaves $f : E \to F$ the map $f(x) : E(x) \to F(x)$ constructed as follows. If $e \in E$, then e may be represented by a section s, i.e. there exists $U \in V_B(x)$ and $s \in E(U)$ such that $e = s_x \in E^U{}_x(s)$. We then put $f(x)(e) = f(U)(s)_x \in F(x)$. It is an easy exercise to verify that this is well-defined and yields a functor $\tau_x : P(B) \to \mathbf{Sets}$, indeed.

(1.28.) A presheaf P on X is said to be a <u>sheaf</u> if the following conditions are satisfied:

(S1) if U is an open subsets of X and $\{U(i); i \in I\}$ is an open covering of U, then for any pair of sections $s, s' \in P(U)$ of P over U, we have $s = s'$, whenever $P^U{}_{U(i)}(s) = P^U{}_{U(i)}(s')$ for all $i \in I$;

(S2) if U is an open subset of X, if $\{U(i); i \in I\}$ is an open covering of U and if for all $i \in I$ there is given some $s_i \in P(U(i))$ with the property that for all $i, j \in I$ we have $P^{U(i)}{}_{U(ij)}(s_i) = P^{U(j)}{}_{U(ij)}(s_j)$, where $U(ij) = U(i) \cap U(j)$, then there exists some $s \in P(U)$ such that $P^U{}_{U(i)}(s) = s_i$ for all $i \in I$.

Note that the presence of (S1) guarantees the element s, whose existence is claimed by (S2), to be unique as such. A presheaf P which satisfies only (S1) is said to be a <u>separated</u> presheaf (or sometimes: a <u>monopresheaf</u>).

We may rephrase axioms (S1) and (S2) together by saying that for any open subset U of X and any open covering $\{U(i); i \in I\}$ of U, the sequence

$$
P(U) \overset{t}{\longrightarrow} \prod_{i \in I} P(U(i)) \underset{d_2}{\overset{d_1}{\rightrightarrows}} \prod_{(i,j) \in I \times I} P(U(i) \cap U(j))
$$

is an equalizer diagram. Here t is given by $t(s)_i = P^U{}_{U(i)}(s)$ for any $i \in I$ and $s \in P(U)$ and d_1 resp. d_2 is defined by putting $d_1((s_i; i \in I))_{(i,j)} = P^{U(i)}{}_{U(ij)}(s_i)$ resp. $d_2((s_i; i \in I))_{(i,j)} = P^{U(j)}{}_{U(ij)}(s_j)$ for any pair of indices $(i, j) \in I \times I$ and $(s_i; i \in I) \in \prod_i P(U(i))$.

If in the above definitions the open sets are only to be taken in some basis B of the topology on X, then we speak of a separated presheaf or a sheaf on B.

(1.29.) Example If E = (E, X, p) is a sheaf space, then ΓE is a sheaf (prove it!). We will show below that all sheaves are essentially of this type, i. e. that a presheaf lies in the image of Γ if and only if it is a sheaf. The presheaves described in (1.19.) and (1.20.) are also easily checked to yield examples of sheaves. The constant presheaf described in (1.18.) is not a sheaf in general, however.

(1.30.) Lemma Let B be a basis of open sets for X, let P be a presheaf on B and Q a separated presheaf on B, then

(1.30.1.) if $f, g : P \to Q$ are presheaf morphisms with the property that for all $x \in X$ the maps $f(x), g(x) : P(x) \to Q(x)$ coincide, then $f = g$;

(1.30.2.) for any open subset $U \in B$ and any $s, s' \in Q(U)$, we have $s = s'$ if and only if $s_x = s'_x$ for all $x \in U$.

Proof (1) Let U be an open subset in B, then we want to show for any $s \in P(U)$ than $f(U)(s) = g(U)(s) \in Q(U)$. Due to the commutativity of the following diagram

$$f(U)$$

$$
\begin{array}{ccc}
P(U) & \longrightarrow & Q(U) \\
\;\;{\scriptstyle P^U_x}\downarrow & & \downarrow{\scriptstyle Q^U_x} \\
P(x) & \longrightarrow & Q(x)
\end{array}
$$

$$f(x)$$

we find that $(f(U)(s))_x = (g(U)(s))_x$, since $f(x)(s_x) = g(x)(s_x)$ by assumption. It thus follows that we may find some $U(x) \in U$ in $V_B(x)$, such that $Q^U_{U(x)}(f(U)(s)) = Q^U_{U(x)}(g(U)(s))$. Since the $U(x)$ cover U, the assertion follows.

(2) One implication is trivial. Conversely, if $s_x = s'_x$ for all $x \in U$, then for any such x we may find an open neighborhood $U(x)$ of x such that $Q^U_{U(x)}(s) = Q^U_{U(x)}(s')$. We find that $s = s'$ exactly as in (1).

(1.31.) Construction We will associate to any presheaf P on some basis B of X a sheaf space $(P^\sim, X, p)$. As a set, we define $P^\sim$ to be the disjoint union of the sets $P(x) = \lim P(U)$, where U runs through $V_B(x)$. This set is endowed with a structure map $p : P^\sim \to X$, which maps any $e \in P^\sim$, which necessarily lies in some $P(x)$ to this unique $x \in X$. For any $s \in P(U)$, where $U \in B$, we define $s^\sim : U \to P^\sim : x \to s_x$ and then clearly $ps^\sim = \mathrm{id}_U$. We put $B = \{s^\sim(U); U \in B, s \in P(U)\}$ and it is easy to see that B is a basis for some topology on $P^\sim$. The maps $s^\sim$ are continuous with respect to this topology (by construction!), so it follows that the $s^\sim$ are sections. Moreover, p is now a local homeomorphism, as one easily checks.
We call $(P^\sim, X, p)$ the <u>sheaf space associated to</u> P. It is easily verified that this yields a (covariant) functor

$$(-)^\sim : \mathbf{P}(B) \to \mathbf{Sh}(X),$$

which to any presheaf P on B associates the sheaf space $(P^\sim, X, p)$ and to any morphism of presheaves $f : P \to Q$ the morphism of sheaf spaces $f^\sim : P^\sim \to Q^\sim$ obtained by glueing together the maps $\tau_x(f) = f(x) : P(x) \to Q(x)$. If V is open in $P^\sim$, then $V = s^\sim(U)$ for some $U \in B$ and some $s \in P(U)$, so $f^\sim(V) = f^\sim(s^\sim(U)) = (f(U)(s))^\sim$, since for any $x \in U$, we have $f^\sim(s^\sim(x)) = f(x)(s_x) = f(U)(s)_x$. Hence $f^\sim$ is open, i.e. a morphism of sheaf spaces, indeed, by the Exercise in (1.13.).
The sheaf $\Gamma(P^\sim, X, p)$ associated to $(P^\sim, X, p)$ is denoted by aP; it is called the <u>associated sheaf</u> of P. Clearly $aP(U) = \Gamma(U, P^\sim)$ for any $U \in \mathrm{Open}(X)$. Since **a** is the composition of Γ and $(-)^\sim$, it follows that we get a covariant functor $a : \mathbf{P}(B) \to \mathbf{P}(B)$. More precisely, if we denote by $\mathbf{S}(B)$ resp. $\mathbf{S}(X)$ the full subcategory of $\mathbf{P}(B)$ resp. $\mathbf{P}(X)$ consisting of sheaves, then we

actually get a functor $a : P(B) \to S(B)$ or $a : P(B) \to S(X)$, since Γ maps sheaf spaces to sheaves!

(1.32.) Example It has already been mentioned that for an arbitrary set A and a basis B for the topology on X, in general the constant presheaf $c_B(A)$ is not a sheaf. So, let us calculate its associated sheaf. The sheaf space $c_B(A)^\sim$ may clearly be identified with A x X and the topology on A x X is determined by the basis $B = \{\{a\} \times U; a \in A, U \in B\}$, i.e. we find the product topology. In other words, the sheaf space $(c_B(A)^\sim, X, p)$ is just the constant sheaf space A_X on X. From (1.6.) it follows that for any $U \in B$ we have $ac_B(A))(U) = A^{c(U)}$, where $c(U)$ is the set of connected components of U. We call $ac_B(A)$ the <u>constant sheaf</u> with fibre A. If B = Open(X), then we denote this by c(X, A), otherwise by $c_B(X, A)$.

(1.33.) Example (the skyscraper sheaf). Let X be a topological space, $y \in X$ and G an arbitrary abelian group. We define a presheaf $\sigma(y, G)$ on X by putting $\sigma(y, G)(U) = G$ if $y \in U$ and $\sigma(y, G)(U) = \{0\}$ otherwise, with obvious restriction morphisms. It is easily seen that $\sigma(y, G)$ is actually a sheaf, which we call the <u>skyscraper sheaf</u> on X with fibre G and support y. It is also clear that $\sigma(y, G)(x) = G$ if and only if x sits in the closure of y, the stalks being $\{0\}$ elsewhere. If X is a Hausdorff space and $G = \mathbf{Z}$, then $\sigma(y, \mathbf{Z})^\sim$ is exactly the skyscraper sheaf space on X with support y and fibre $\mathbf{Z}$, described in (1.7.)

(1.34.) Proposition Let B be a basis of open subsets for the topology on X, then

(1.34.1.) there exists a natural transformation $\eta : \mathrm{id}_{P(B)} \to a$;

(1.34.2.) a presheaf P is separated if and only if $\eta_P : P \to aP$ is a monomorphism;

(1.34.3.) a presheaf P is a sheaf if and only if $\eta_P : P \to aP$ is an isomorphism.

Proof Define η by putting for any $U \in B$ and any $P \in \mathbf{P}(B)$

$$\eta_P(U) : P(U) \to \mathbf{a}P(U) : s \to s^\sim.$$

We leave it to the reader to verify that this defines a natural transformation η indeed.

To prove the second claim, first note that a presheaf morphism $f : P \to Q$ is monic if and only if for each $U \in B$ the map $f(U) : P(U) \to Q(U)$ is injective. Now, assume that P is separated and let $U \in B$ resp $s, t \in P(U)$ be such that $\eta_P(U)(s) = \eta_P(U)(t)$, i.e. $s^\sim = t^\sim$, then $s_x = t_x$ for all x in U, hence $s = t$ by (1.30.2.), so η_P is monic.

Conversely, assume η_P is monic, let $\{U(i); i \in I\}$ be an open covering in B and take $s, t \in P(U)$ such that $P^U_{U(i)}(s) = P^U_{U(i)}(t)$ for all $i \in I$. Since $(\mathbf{a}P)^U_{U(i)}\eta_P(U) = \eta_P(U(i))P^U_{U(i)}$, it follows that $(\mathbf{a}P)^U_{U(i)}(\eta_P(U)(s)) = (\mathbf{a}P)^U_{U(i)}(\eta_P(U)(t))$ for all $i \in I$, hence $\eta_P(U)(s) = \eta_P(U)(t)$, since $\mathbf{a}P$ is a sheaf, so $s = t$, since $\eta_P(U)$ is monic. Hence P is separated, as claimed.

Finally, to check the last assertion, it is clear that we only have to prove that for any $P \in \mathbf{S}(B)$ and any $U \in B$, the map $\eta_P(U) : P(U) \to \mathbf{a}P(U)$ is bijective. Due to (2), it is certainly injective. To prove its surjectivity, take $t \in \Gamma(U, P^\sim)$, then $t(U)$ is open in $P^\sim$. For all $x \in U$, the element $t(x) \in (P^\sim)_x = P(x)$ possesses a basic open neighborhood $(s^x)^\sim(U(x))$ inside of $t(U)$, with $U(x) \in V_B(x)$ and $s^x \in P(U(x))$. This means that the s^x satisfy the glueing property for sheaves by (1.30.2.), hence there exists an $s \in P(U)$ with the property that $P^U_{U(x)}(s) = s^x$ for all x in U, so $s^\sim(x) = (s^x)_x = t(x)$, for all $x \in U$ and so $\eta_P(U)(s) = t$, proving the surjectivity of $\eta_P(U)$.

(1.35.) Proposition For any sheaf space $\mathbf{E} = (E, X, p)$ the sheaf spaces E and $\Gamma(\,\text{-}\,, \mathbf{E})^\sim$ are isomorphic.

Proof This means that there is an open bijection $\phi : E \to \Gamma(\,\text{-}\,, \mathbf{E})^\sim$ such that the following diagram commutes

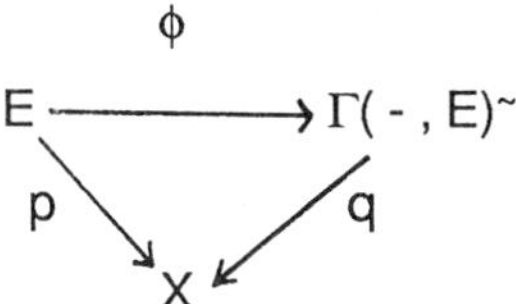

Indeed, ϕ is then automatically a homeomorphism, by (1.13.). Let $x \in X$, then by (1.12.) there exists a bijection between $p^{-1}(x)$ and the stalk $\Gamma(\, , E)(x)$, i.e. the fibre $q^{-1}(x)$ in $\Gamma(- , E)^{\sim}$. These bijections glue together to a bijection $\phi : E \to \Gamma(- , E)^{\sim}$, which makes the diagram commute. If U is open in X and if $s \in \Gamma(U, E)$, then $f(s(U)) = s^{\sim}(U)$, so f is open indeed.

(1.36.) Proposition The functor $\mathbf{a} : \mathbf{P}(B) \to \mathbf{S}(B)$ is left adloint to the canonical inclusion $i : \mathbf{S}(B) \to \mathbf{P}(B)$.

Note It will be clear from the proof below that the functors $\mathbf{a} : \mathbf{P}(B) \to \mathbf{S}(X)$ and $i : \mathbf{S}(X) \to \mathbf{P}(B)$ are also adjoint to each other, ...

Proof We will prove that for any presheaf P on B and any sheaf S on B, a presheaf morphism $f : P \to S$ factorizes uniquely through aP as

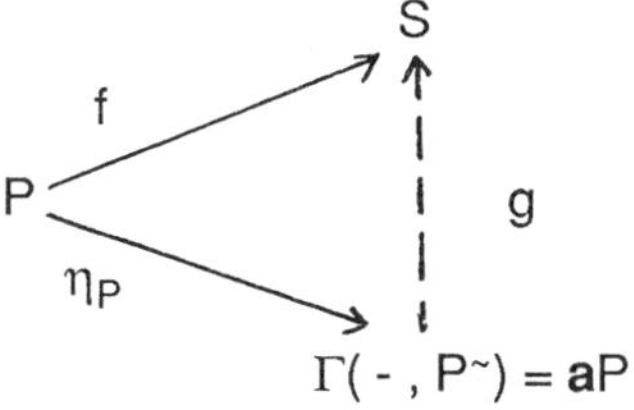

Indeed, if such a g exists, then the stalk morphisms are determined by

$$g(x) : (aP)(x) \xrightarrow{\;\tau_x(\eta_P)^{-1}\;} P(x) \xrightarrow{\;\tau_x(f)\;} S(x)$$

hence g is necessarily unique. On the other hand, if $f : P \to S$ is a presheaf morphism, then the functor $(-)^{\sim} : \mathbf{P}(B) \to \mathbf{Sh}(X)$ yields a map $f^{\sim} :$

$P^{\sim} \to S^{\sim}$, hence a morphism $\Gamma(- , f^{\sim}) : \Gamma(- , P^{\sim}) = \mathbf{a}P \to \mathbf{a}S = \Gamma(- , S^{\sim})$. Let $g = (\eta_S)^{-1}\Gamma(- , f^{\sim})$, then an easy calculation shows that $g\eta_P = f$, as claimed. We leave it as an exercise to the reader to verify that the bijection

$$\tau_{P, S} : [P, iS]_{\mathbf{P}(B)} \to [\mathbf{a}P, S]_{\mathbf{S}(B)} : f \to g$$

this construction defines, is functorial in P and S, showing that $\mathbf{a}$ and i are adjoint to each other.

2   SHEAVES OF ABELIAN GROUPS

(2.1.) In this Section we will endow sheaves and sheaf spaces with more structure. We will keep the notations and terminology of the previous Section. In particular, B will always denote a basis of open subsets for the topology on some fixed topological space X.

(2.2.) A (pre)sheaf P on B is said to be a <u>(pre)sheaf of abelian groups</u> if $P(U)$ is an abelian group for every U in B and if for every $V \subset U$ in B the restriction morphism $P^U_V : P(U) \to P(V)$ is a group homomorphism.

If we start from a presheaf of abelian groups P, it thus follows that for each x in X the stalk $p^{-1}(x) = P(x) = \lim P(U)$ of the associated sheaf space $(P^{\sim}, X, p)$ is an abelian group.

Now, let (E, X, p) be an arbitrary sheaf space with this property and denote by $E \times_X E$ the subspace of $E \times E$ consisting of all (e, e') with $p(e) = p(e')$ (with the induced topology), then we claim that $\Gamma(- , E)$ is a sheaf of abelian groups if and only if the map $m : E \times_X E \to E : (e, e') \to e - e'$ is continuous. Indeed, assume that for all open subsets U of X the set $\Gamma(U, E)$ is an abelian group (under pointwise addition), then we want to verify whether for every $s \in \Gamma(U, E)$ the set $m^{-1}(s(U))$ is open. Now, from $(e, e') \in m^{-1}(s(U))$, it follows that $s(x) = e - e'$, with $x = p(e) = p(e')$. Take W open in X and $t, t' \in \Gamma(W, E)$ such that $t(x) = e$ and $t'(x) = e'$, then $(t - t')(x) = s(x)$, so for some open subset V of $W \cap U$ in $V(x)$, it follows that $E^W_V(t - t') = E^U_V(s)$, hence (e, e') possesses the open neighborhood $(t(V) \times t'(V)) \cap (E \times_X E)$ within $m^{-1}(s(U))$.

Conversely, if $s, t \in \Gamma(U, E)$, then we may factorize $s - t : U \to E$ in the following way

$$
\begin{array}{ccc}
 & (s, t) & m \\
U \longrightarrow & E \times_X E & \longrightarrow E \\
x \longmapsto & (s(x), t(x)) & \longmapsto s(x) - t(x),
\end{array}
$$

hence it is continuous and s - t $\in$ Γ(U, E). The rest is now easy.

The foregoing of course also works on some basis B of the topology on X. A sheaf space (E, X, p) satisfying the foregoing conditions is called a <u>sheaf space of abelian groups</u>. It is clear that there is a one-to-one correspondence between sheaf spaces of abelian groups and sheaves of abelian groups.

(2.3.) A <u>morphism of (pre)sheaves of abelian groups</u> f : P $\to$ Q is just a morphism of presheaves such that for each U in B the map f(U) : P(U) $\to$ Q(U) is a group homomorphism. These morphisms being composed in the obvious way, this yields categories $\underline{\mathbf{S}}$(B) $\subset$ $\underline{\mathbf{P}}$(B), the category of (pre)sheaves of abelian groups on B. As usually, if B = Open(X), then we write $\underline{\mathbf{S}}$(X) resp. $\underline{\mathbf{P}}$(X) for the corresponding categories.

A <u>morphism of sheaf spaces of abelian groups</u> f : **E** = (E, X, p) $\to$ **F** = (F, X, q) is just a morphism of sheaf spaces in the usual sense, such that for each x $\in$ X the stalk map f_x : $p^{-1}(x) = E_x \to F_x = q^{-1}(x)$ is a group homomorphism. These morphisms may also be composed, and this yields the category $\underline{\mathbf{Sh}}$(X) of sheaf spaces of abelian groups on X. In general however, we will identify the categories $\underline{\mathbf{S}}$(X) and $\underline{\mathbf{Sh}}$(X), through the equivalences (-)$^\sim$: $\underline{\mathbf{S}}$(X) $\approx$ $\underline{\mathbf{Sh}}$(X) and Γ : $\underline{\mathbf{Sh}}$(X) $\approx$ $\underline{\mathbf{S}}$(X) (cf. (1.34.) and (1.35.)!). This permits us to interchange notations as P(U) and Γ(U, P) without ambiguity. As a matter of fact, even if P is only a presheaf of abelian groups on B, we will sometimes write Γ(U, P) instead of P(U), for any U in B. Note also that **a** restricts to a functor **a** : $\underline{\mathbf{P}}$(B) $\to$ $\underline{\mathbf{S}}$(B).

If $\underline{\mathbf{C}}$ denotes one of the categories $\underline{\mathbf{S}}$(X), $\underline{\mathbf{Sh}}$(X), $\underline{\mathbf{P}}$(X), $\underline{\mathbf{P}}$(B) or $\underline{\mathbf{S}}$(B), then for any U in B we thus have a functor

$$\Gamma(U, -) : \underline{\mathbf{C}} \to \textbf{Abelian Groups},$$

which to any morphism f : F $\to$ G in $\underline{\mathbf{C}}$ associates the morphism Γ(U, f) :

$\Gamma(U, F) \to \Gamma(U, G)$. As we will see, sheaf theory is strongly concerned with the exactness properties of this functor.

(2.4.) Of course, we will have to put more structure on the category $\underline{C}$ first. Let us start by making $\underline{P}(B)$ and $\underline{S}(B)$ into additive categories. If F and G are in $\underline{C}$, then for any f, g $\in$ $[F, G]_{\underline{C}}$, we construct the sum f + g $\in$ $[F, G]_{\underline{C}}$, by putting $(f + g)(U) : F(U) \to G(U) : s \to f(U)(s) + g(U)(s)$. There is an obvious zero map 0 $\in$ $[F, G]_{\underline{C}}$ and in this way the sets $[F, G]_{\underline{C}}$ become abelian groups, such that the composition of morphisms in $\underline{C}$ is bilinear. The zero object in $\underline{C}$ is given by $0(U) = \{0\}$ for all U in B, with obvious restriction maps, and a fairly trivial verification thus shows $\underline{C}$ to be an abelian category, indeed.

(2.5.) Let $f : F \to G$ be a morphism in $\underline{P}(B)$, then for any U in B, we put $K(U) = Ker(f(U)) = \{s \in F(U); f(U)(s) = 0\}$. It is clear that the restriction maps F^U_V induce restriction maps $K^U_V : K(U) \to K(V)$, hence this construction defines a presheaf of abelian groups K on B, which we call the <u>kernel</u> of f and which we denote by Ker(f). This presheaf is a <u>subpresheaf</u> of F in the sense that $Ker(f)(U) \subset F(U)$ for all U in B. We thus write : $Ker(f) \subset F$.

Exercise If $f : F \to G$ is a morphism of presheaves of abelian groups and if F is a sheaf resp. G a separated presheaf, then Ker(f) is a sheaf.
It follows in particular that for any morphism $f : F \to G$ in $\underline{S}(B)$, we have $Ker(f) \in \underline{S}(B)$.

(2.6.) Proposition Kernels exist in $\underline{P}(B)$ and $\underline{S}(B)$.
Proof It suffices to show that with $\underline{C} = \underline{P}(B)$ or $\underline{S}(B)$, for any $f : F \to G$ in $\underline{C}$, the object Ker(f) satisfies the following universal property : if H $\in$ $\underline{C}$ and g $\in$ $[H, F]$ is such that fg = 0, then g factorizes uniquely through Ker(f):

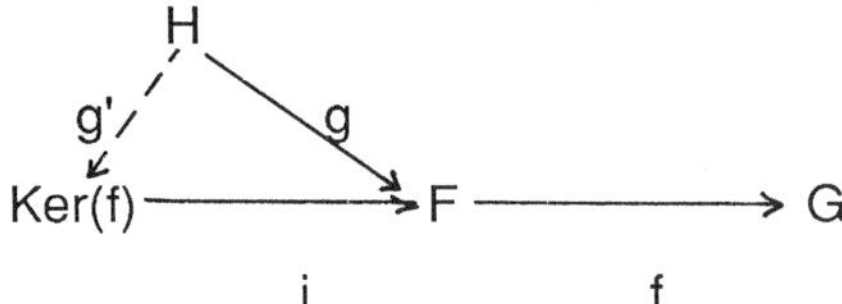

The proof of this follows easily from the very construction of Ker(f), so we leave it as an easy exercise to the reader. Note also that this universal property may be viewed as an alternative definition of Ker(f).

(2.7.) From the exactness properties of inductive limits, it follows that for any x in X and any $f : F \to G$ in $\underline{C}$, we have $\mathrm{Ker}(f)_x = \mathrm{Ker}(f_x)$, where f_x is the stalk morphism at x. In particular, one easily checks that f is a monomorphism in $\underline{S}(B)$ if and only if for all x in X the map f_x is injective or equivalently, if f(U) is injective for all U in B or if Ker(f) = 0. The last two statements are equivalent to f being a monomorphism in $\underline{P}(B)$.

(2.8.) Proposition Cokernels exist in $\underline{P}(B)$.
Proof Let $f : F \to G$ be a morphism in $\underline{P}(B)$ and write C(U) = G(U)/Im(f(U)) for any $U \in B$. If $V \subset U$, then we obtain an induced map $C^U{}_V : C(U) \to C(V)$ and it is easy to see that this defines a presheaf of abelian groups on B, which we denote by $\mathrm{Coker}_p(f)$. The quotient maps $G(U) \to C(U)$ define a morphism $\pi : G \to \mathrm{Coker}_p(f)$ in $\underline{P}(B)$, and argueing as in (2.6.) one may verify that this a cokernel for f in $\underline{P}(B)$, indeed.

(2.9.) As for kernels, one easily shows that $\mathrm{Coker}_p(f) = 0$ is equivalent to f being an epimorphism in $\underline{P}(B)$ or to f(U) being surjective for all U in B. However, for sheaves of abelian groups, the situation is much more delicate here, since in general we do not have that for any morphism $f : F \to G$ in $\underline{S}(B)$ the presheaf cokernel $\mathrm{Coker}_p(f)$ lies in $\underline{S}(B)$ too.

(2.10.) Example Consider the topological space $[0, 1] = X$, with non trivial open sets $[0, 1[= U(1)$, $]0, 1] = U(2)$ and $]0, 1[= V$. All of these being connected, the constant presheaf F with fibre **Z** on X is a sheaf of abelian groups. Consider the sheaf G on X defined as follows : we put $G(X) = \mathbf{Z} \times \mathbf{Z}$, $G(U(i)) = \mathbf{Z}$ and $G(V) = \{0\}$. The only non trivial restriction maps are the maps $G^X_{U(i)} : G(X) = \mathbf{Z} \times \mathbf{Z} \to \mathbf{Z} = G(U(i))$, the projection on the i-th factor, the others being the identity or the zero map.

We now define a morphism $f : F \to G$ by putting over X the diagonal map $\mathbf{Z} \to \mathbf{Z} \times \mathbf{Z}$, over each $U(i)$ the identity map $\mathbf{Z} \to \mathbf{Z}$ and over V the zero map $\mathbf{Z} \to \{0\}$. It is then easy to see that for all open U in X we have $\mathrm{Coker}_p(f)(U) = \{0\}$, except for $U = X$, where $\mathrm{Coker}_p(f)(X) = \mathbf{Z}$. Hence $\mathrm{Coker}_p(f)$ is certainly not a sheaf. Note also that $G_x = \{0\}$ for each x in X, except for $x = 0$ or 1, where $G_x = \mathbf{Z}$. In particular, it follows that $f_x : F_x \to G_x$ is surjective for all x in X.

(2.11.) Proposition Cokernels exist in $\underline{S}(B)$.

Proof Let $f : F \to G$ denote a morphism in $\underline{S}(B)$, then we define the sheaf cokernel $\mathrm{Coker}_s(f)$ to be the sheaf $\mathbf{a}\mathrm{Coker}_p(f)$, the sheaf associated to the presheaf cokernel of f. Clearly $\mathrm{Coker}_s(f)$ is a sheaf of abelian groups and there is a canonical morphism $\pi_s : G \to \mathrm{Coker}_s(f)$ such that $\pi_s f = 0$. Moreover, $\mathrm{Coker}_s(f)$ satisfies the following universal property: if H is a sheaf of abelian groups on B and $g \in [G, H]_{\underline{S}(B)}$ is such that $gf = 0$, then g factorizes uniquely through $\mathrm{Coker}_s(f)$:

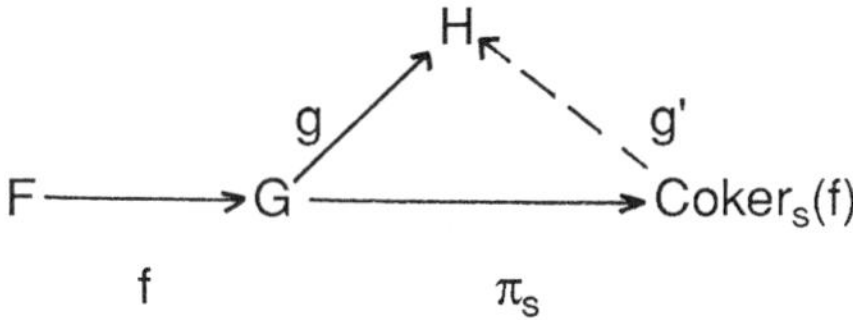

Indeed, this follows immediately from the corresponding universal property on the presheaf level. Moreover, g' is necessarily unique as such, for if not, there is another morphism $g'' : \mathrm{Coker}_s(f) \to H$ with $g''\pi_s = g'\pi_s = g$, hence by (1.30.) there exists at least one x in X such that g'_x and g''_x do not coincide. But this is impossible, since $\mathrm{Coker}_s(f)_x = \mathrm{Coker}_p(f)_x = \mathrm{Coker}(f_x)$ and since the universal property of cokernels in the category of abelian groups implies that there is a unique morphism (g'_x !) such that $g'_x\pi_x = g_x$, where $\pi_x = (\pi_s)_x$ is the canonical map $G_x \to \mathrm{Coker}(f_x)$. This proves the assertion.

(2.12.) It follows from this that a morphism $f : F \to G$ in $\underline{S}(B)$ is an epimorphism if and only if $\mathrm{Coker}_s(f) = 0$ or if $f_x : F_x \to G_x$ is surjective for all x in X. Note however, in contrast with the presheaf situation, that an epimorphism $f : F \to G$ in $\underline{S}(B)$ does not necessarily yield a surjective morphism $f(U) : F(U) \to G(U)$ for every U in B. Indeed, (2.10.) yields a counterexample, as in this case $f(X)$ is certainly not surjective (it is the diagonal map $\mathbf{Z} \to \mathbf{Z} \times \mathbf{Z}$), whereas f_x is surjective for all $x \in X$.

(2.13.) Corollary A morphism $f : F \to G$ in $\underline{P}(B)$ is an isomorphism if and only if it is a monomorphism and a presheaf epimorphism or equivalently if $f(U)$ is bijective for all $U \in B$. If f is a morphism in $\underline{S}(X)$, this is also equivalent to f being a monomorphism and a sheaf epimorphism or to f_x being a bijection for all x in X.

(2.14.) The same problems occur with images : if $f : F \to G$ is a morphism in $\underline{P}(B)$, then $\mathrm{Im}_p(f)$, defined by $\mathrm{Im}_p(f)(U) = \mathrm{Im}(f(U)) \subset G(U)$ for each U in B enjoys the universal properties of a genuine image in the category $\underline{P}(B)$, but in general for F and G in $\underline{S}(B)$ we do not have $\mathrm{Im}_p(f) \in \underline{S}(B)$. However, since $\mathrm{Im}_p(f) \subset G$ and G is a sheaf, it follows that $\mathrm{Im}_p(f)$ is a separated presheaf on B. If we write $\mathrm{Im}_s(f) = \mathbf{a}\mathrm{Im}_p(f)$, then we thus have $\mathrm{Im}_p(f) \subset \mathrm{Im}_s(f)$ and one may verify that this "sheaf image" enjoys all of the universal properties of an image in $\underline{S}(B)$. Moreover, it is fairly straightforward to verify that $\mathrm{Im}_p(f)_x = \mathrm{Im}_s(f)_x = \mathrm{Im}(f_x)$ for all x in X. We now leave it as an exercise to the reader to prove the following result, which strengthens the remarks made in (2.4.).

(2.15.) Proposition The categories $\underline{P}(B)$ and $\underline{S}(B)$ are abelian.

(2.16.) As usually, a sequence of morphisms $F \to^f G \to^g H$ in $\underline{P}(B)$ resp. $\underline{S}(B)$ is said to be <u>exact</u> if $\text{Im}(f) = \text{Ker}(g)$, where, as we have pointed out above, $\text{Im}(f) = \text{Im}_p(f)$ if we work in $\underline{P}(B)$ and $\text{Im}(f) = \text{Im}_s(f)$ if we work in $\underline{S}(B)$. We call a sequence $0 \to F \to^f G \to^g H \to 0$ in $\underline{P}(B)$ or $\underline{S}(B)$ a <u>short exact sequence</u>, if each subsequence consisting of two consecutive maps yields an exact sequence. In particular, this says that f is monic, g epic and that $\text{Im}(f) = \text{Ker}(g)$. From the foregoing, it follows that $\mathbf{a} : \underline{P}(B) \to \underline{S}(B)$ is an exact functor, i.e. that it respects exact sequences.

Due to the fact that kernels and cokernels behave nicely in $\underline{P}(B)$, it follows that the functor $\Gamma(U, \text{ - }) : \underline{P}(B) \to \textbf{abelian groups}$ is exact i.e. if $0 \to E' \to E \to E'' \to 0$ is a short exact sequence of presheaves of abelian groups, then the induced sequence of abelian groups

$$0 \to E'(U) \to E(U) \to E''(U) \to 0$$

is exact as well.

On the other hand, if $0 \to E' \to E \to E'' \to 0$ is exact in $\underline{S}(B)$, then it only induces an exact sequence of abelian groups $0 \to E'(U) \to E(U) \to E''(U)$, i.e. $\Gamma(U, \text{ - }) : \underline{S}(B) \to \textbf{abelian groups}$ is only <u>left</u> exact!

(2.17.) Since an exact sequence of sheaves as above only induces a short exact sequence $0 \to E'(U) \to E(U) \to E''(U)$, one is clearly interested in the cokernel of $E(U) \to E''(U)$. It appears that this cokernel may be described by some cohomology group. A lot of excellent treatments of sheaf cohomology may be found in the literature, so we will not give any details on the subject here. As a matter of fact, in this text, we will always be able to restrict to Cech cohomology groups, so we have preferred not to talk about general (Grothendieck) cohomology at all. Yet, although Cech cohomology is much easier to calculate (especially since we will mainly have to deal with H^1!) the reader should realize that Cech cohomology is not the "real" cohomology, this role being reserved to Grothendieck cohomology... However, for H^1 no harm is done : the Cech H^1 and the Grothendieck H^1 coincide (cf. [Ha1] for example).

(2.18.) Let us start with some definitions. Let E be a sheaf of abelian groups on X and let $U = (U(i); i \in I)$ be an open covering of X. To avoid set-theoretical complications, it is sometimes handy to assume that $I = X$ and to let $U(x) \in V_X(x)$ for any x in X. Instead of $U(i_0) \cap ... \cap U(i_p)$, where $\{i_0, ... , i_p\} \subset I$, we will write $U(i_0 ... i_p)$. This allows us to define a complex $C^*(U, E)$ of abelian groups as follows. For any positive integer p we put $C^p(U, E) = \Pi E(U(i_0 ... i_p)$ and $d^p : C^p(U, E) \to C^{p+1}(U, E)$ is given by

$$d^p((s(i_0, ..., i_p); (i_0, ..., i_p) \in I^{p+1})) = (t(i_0, ..., i_{p+1}); (i_0, ..., i_{p+1}) \in I^{p+2}),$$

where $s(i_0, ..., i_p) \in E(U(i_0, ..., i_p))$ for every $(i_0, ..., i_p) \in I^{p+1}$ and where the element $t(i_0, ..., i_{p+1})$ is given by

$$t(i_0, ..., i_{p+1}) = \Sigma_r (-1)^r s(i_0, ..., (i_r)^\wedge, ..., i_{p+1})|U(i_0 ... i_{p+1}),$$

where r runs from 0 to p+1 and $(i_r)^\wedge$ means : omit i_r. One easily verifies that $d^{p+1}d^p = 0$ for every p, hence $(C^*(U, E), d^*)$ is a complex, indeed. By convention we define $d^{-1} : \{0\} \to C^0(U, E)$ to be the trivial map. The p-th cohomology group, defined by $H^p(U, E) = \mathrm{Ker}(d^p)/\mathrm{Im}(d^{p-1})$ is called the p-th <u>Cech cohomology group</u> of E with respect to U.

(2.19.) A <u>refinement</u> of the covering U is a covering $V = (V(j); j \in J)$, together with a map $u : J \to I$, such that for each $j \in J$ we have $V(j) \subset$

$U(u(j))$. If V is a refinement of U, then the map u induces a morphism of complexes

$$u^* : (C^*(U, E), d^*_U) \to (C^*(V, E), d^*_V),$$

the d^*_U resp. d^*_V being the boundary maps for U resp. V, hence for each i a morphism $\tau^p(u) : H^p(U, E) \to H^p(V, E)$ on the cohomology groups. We leave it as an exercise to the reader to show that the map $\tau^p(u)$ is actually independent of u. **Hint**: for any pair of refinement maps $u, v : J \to I$, construct a homotopy between the morphisms of complexes $u^*, v^* : (C^*(U, E), d^*_U) \to (C^*(V, E), d^*_V)$. General homological algebra then yields the assertion.

The class of coverings being partially ordered (under refinement), we may calculate the inductive limit over all coverings and this yields the groups $H^p(X, E) = \lim H^p(U, E)$, the p-th <u>Cech cohomology group</u> of E over X. Note that there is no ambiguity in notation, since we will not use Grothendieck cohomology in this text.

(**Note** : There is actually a small set-theoretic difficulty here, since the class of all open coverings of X is not a set in general. This problem may be avoided, however, e.g. by taking coverings of X indexed by X itself or chosen in some sufficiently large set. We will not go into this here, since in practice all our coverings may be chosen to be indexed by (subsets of) X!).

(**2.20.**) **Proposition** For any sheaf of abelian groups E on X, there is a canonical isomorphism $H^0(X, E) = \Gamma(X, E) = E(X)$.

Proof In fact, one may show that $H^0(X, E) = E(X)$ for any open covering $U = (U(i); i \in I)$ of X. Indeed, we have $H^0(U, E) = \mathrm{Ker}(d^0 : C^0(U, E) \to C^1(U, E))$, where $d^0 : \prod_i E(U(i)) \to \prod_{(i, j)} E(U(ij))$ is given by $d^0((s(i); i \in I)) =$

$(s(i, j) \in I \times I)$, with $s(i, j) = s(j)|U(ij) - s(i)|U(ij)$ for any pair of indices i, j in I. The sheaf axioms then imply that $H^0(U, E) = E(X)$, indeed.

(2.21.) Proposition For any exact sequence of sheaves of abelian groups

$$0 \to E' \xrightarrow{f} E \xrightarrow{g} E'' \to 0,$$

there is an exact sequence of Cech cohomology groups

$$0 \to E'(X) \xrightarrow{f(X)} E(X) \xrightarrow{g(X)} E''(X) \xrightarrow{\delta} H^1(X, E') \xrightarrow{f^1(X)} H^1(X, E) \xrightarrow{g^1(X)} H^1(X, E'')$$

Proof Let us first define $\delta : E''(X) \to H^1(X, E')$. Pick $s'' \in E''(X)$, then for all x in X, we may find some open neighborhood $W(x)$ of x and some s^x in $E(W(x))$, such that $g_x((s^x)_x) = s''_x$ (since g_x is surjective), hence such that $g(W(x))(s^x)_x = s''_x$. It follows that for some neighborhood $U(x) \subset W(x)$ of x we have $s''|U(x) = g(W(x))(s^x)|U(x) = g(U(x))(s^x|U(x))$. Let $t^x = s^x|U(x) \subset E(U(x))$ and write $U(xy)$ for $U(x) \cap U(y)$, then we have $g(U(xy))(t^x|U(xy)) = g(U(xy))(s^x|U(xy)) = g(U(x))(s|U(x))|U(xy) = s''|U(xy)$, hence $g(U(xy))(t^x|U(xy)) = g(U(xy))(t^y|U(xy))$, i.e. $t^x|U(xy) - t^y|U(xy) \in \mathrm{Ker}(g(U(xy))) = \mathrm{Im}(f(U(xy)))$. We may thus find a unique $t(x, y) \in E'(U(xy))$ such that $f(U(xy))(t(x, y)) = t^x|U(xy) - t^y|U(xy)$. If we let $U = (U(x); x \in X)$, then this yields an element $t = (t(x, y); (x, y) \in X \times X) \in C^1(U, E')$, and one easily checks that $t \in \mathrm{Ker}(d^1)$. We thus get an element $\alpha \in H^1(X, E')$ through the map $\mathrm{Ker}(d^1) \to \mathrm{Ker}(d^1)/\mathrm{Im}(d^0) = H^1(U, E') \to \lim H^1(U, E') = H^1(X, E')$. We leave it to the reader to verify that α is well defined, i.e. that it does not depend upon the choices made. This proves that δ is also well defined and it is easy to see that δ is a group homomorphism.

By its very definition, we have $\delta g(X) = 0$. On the other hand, let $\delta(s'') = 0$ for some $s'' \in E''(X)$, then we may find some covering $U = (U(x); x \in X)$ of X together with elements $t^x \in E(U(x))$ such that $(t(x, y); (x, y) \in X \times X)$ constructed as before, yields the zero class in $H^1(U, E')$, i.e. $(t(x, y)); (x, y) \in X \times X) \in \text{Im}(d^0)$. We may thus find $(u^x; x \in X)$ with $u^x \in E'(U(x))$ and with $u^x|U(xy) - u^y|U(xy) = t(x, y)$ for each $x, y \in X$, hence we may find some r in $E(X)$ such that $t^x - f(U(x))(u^x) = r|U(x)$ for each x in X. But $g(U(x))(t^x) = s''|U(x)$ and $u^x \in E'(U(x)) = \text{Ker}(g(U(x)))$, so $g(X)(r)|U(x) = g(U(x))(r|U(x)) = g(U(x))(t^x - f(U(x))(u^x)) = s''|U(x)$. Since this holds for all x in X, this proves that $s'' = g(X)(r)$, i.e. $s'' \in \text{Im}(g(X))$. So the sequence is defined and exact up to δ.

The maps $f^1(X)$ and $g^1(X)$ are defined through the maps $H^1(U, E') \to H^1(U, E)$ and $H^1(U, E) \to H^1(U, E'')$ for any open covering U, which in turn are defined through the obvious maps $C^1(U, E') \to C^1(U, E)$ resp. $C^1(U, E) \to C^1(U, E'')$. We leave it to the reader to verify that $\text{Im}(f^1(X)) = \text{Ker}(g^1(X))$ (easy !).

It remains to prove that the sequence is exact at $H^1(X, E')$. First, let $s'' \in E''(X)$ and choose an open covering $U = (U(x); x \in X)$ with elements $t^x \in E(U(x))$ having the property that $s''|(x) = g(U(x))(t^x)$ and such that $\delta(s'')$ is given by the element in $H^1(U, E')$ represented by $(t(x, y); (x, y) \in X \times X)$ with $f(U(xy))(t(x, y)) = t^x|U(xy) - t^y|U(xy)$. The map $f^1(X)$ is represented over U by $f_U : C^1(U, E') \to C^1(U, E)$ defined by

$$f_U((t(x, y); (x, y) \in X \times X)) = (f(U(xy))(t(x, y)); (x, y) \in X \times X).$$

But by the choice of the $t(x, y)$, it follows that $f_U((t(x, y); (x, y) \in X \times X) \in \text{Im}(d^0)$, i.e. $f^1(X)\delta = 0$.

Conversely, assume that $f^1(X)(\alpha) = 0$ for some $\alpha \in H^1(X, E')$. Represent α by some $\sigma = (s(x, y); (x, y) \in X \times X) \in C^1(U, E')$ for some suitable covering $U = (U(x); x \in X)$ of X, then, upon refining U if necessary, we may assume that $f_U(s) = 0$ in $H^1(U, E)$, i.e. $f_U(\sigma) \in \text{Im}(d^0)$. We may thus find a family $(t^x; x \in X) \in C^0(U, E)$ such that for all x, y in X one has $f(U(xy))(s(x, y)) = t^x|U(xy) - t^y|U(xy)$. But then $g(U(x))(t^x)|U(xy) = g(U(y))(t^y)|U(xy)$, so there exists some s" $\in$ E"(X) with $s"|U(x) = g(U(x))(t^x)$ for all $x \in X$. It follows that $\alpha = \delta(s")$, i.e. $\text{Im}(\delta) = \text{Ker}(f^1(X))$.

Note Using Grothendieck cohomology one may extend the above exact sequence to higher cohomology groups.

(2.22.) In the next paragraphs, we will always work over Open(X). The reader is invited to adapt the definitions and results below to sheaves over a basis B of open sets of X.

A sheaf E is said to be <u>flabby</u> if its restriction morphisms E^U_V are surjective for all $V \subset U$. Examples of flabby sheaves abound. In fact there is a functor $\gamma: \mathbf{S}(X) \to \mathbf{S}(X)$ (restricting to $\gamma : \underline{\mathbf{S}}(X) \to \underline{\mathbf{S}}(X)$!) which to any sheaf E on X associates a flabby sheaf γE endowed with a monomorphism $E \subset \gamma E$ (actually, there exists a natural transformation $\text{id}_{\mathbf{S}(X)} \to \gamma$!). Indeed let $E \in \mathbf{S}(X)$, then for any open subset U of X we put $\gamma E(U) = \Pi E_x$, where the product is taken over all $x \in U$. With the obvious restriction morphisms, this is easily seen to yield (in a functorial way) a flabby sheaf. The morphism $j_E : E \to \gamma E$ is defined by putting for each open subset U of X

$$j_E(U) : E(U) \to \gamma E(U) = \Pi E_x : s \to (s_x; x \in U).$$

It follows from (1.30.2.) that this is injective, indeed.

(2.23.) Actually, the functor γ may be generalized as follows: we claim that for any family $(E^x; x \in X)$ of sets we may construct a sheaf $E = \gamma(E^x; x \in X)$ such that for each sheaf F there is a natural bijection

$$[F, E]_{\mathbf{S}(X)} = \Pi_x [F_x, E^x]_{\mathbf{Sets}}.$$

The result remains valid if we work with (sheaves of) abelian groups.

Of course, E is defined by $E(U) = \Pi E^x$, where x runs through U, for any open subset U of X. The bijection between the morphism sets is determined as follows: if $f : F \to E$ is a sheaf morphism, then for each x in X and each open neighborhood U of x, it yields a map $F(U) \to E(U) = \Pi E^x \to E^x$ (the second map is the projection), and since these form an inductive system, we obtain a map $F_x \to E^x$, indeed.

(2.24.) Corollary The category $\underline{S}(X)$ has enough injectives.

Proof This means that every sheaf of abelian groups E may be embedded in a sheaf of abelian groups I which is an injective object in $\underline{S}(X)$. But, it is well known that each of the abelian groups E_x may be be embedded in an injective (or divisible) abelian group, I^x say. The foregoing construction then yields a sheaf of abelian groups $I = \gamma((I^x; x \in X))$ together with a monomorphism $E \to I$ and another application of the construction shows that I is then injective, indeed.

(2.25.) Actually, it should come as no surprise to us that the injective sheaves constructed in (2.24.) are flabby, since we will see below that injectives in $\underline{S}(X)$ are always flabby.

Let U be an open subset of X, then we may construct two functors $(-)^X$ and $(-)|U$ as follows:

(2.25.1.) Extension by zero For any sheaf of abelian groups E on U, we construct a sheaf of abelian groups E^X on X, the extension of E by zero, by putting $E^X(V) = E(V)$ if $V \subset U$ and $E^X(V) = \{0\}$ otherwise, the restriction maps being the obvious ones.

(2.25.2.) Restriction For any sheaf of abelian groups F on X, the restriction of F to U, denoted by $F|U$, by putting $(F|U)(V) = F(V)$ for every open subset V of U. Clearly $F|U$ is a sheaf of abelian groups on U. As a matter of fact, if F is represented by the sheaf space (F, X, p), then $F|U$ is represented by the sheaf space $(p^{-1}(U) = F|U, U, p|U)$.

One easily sees that $(-)^X$ and $(-)|U$ are functors. In fact:

(2.26.) Proposition The functors $(-)^X : \underline{S}(U) \to \underline{S}(X)$ and $(-)|U : \underline{S}(X) \to \underline{S}(U)$ are adjoint to each other.

Proof We have to verify that for any $E \in \underline{S}(U)$ and any $F \in \underline{S}(X)$, there is a (functorial) bijection

$$\tau = \tau_{E,F} : [E^X, F]_{\underline{S}(X)} \to [E, F|U]_{\underline{S}(U)}.$$

Let $f : E^X \to F$ be a morphism in $\underline{S}(X)$, then $\tau(f) : E \to F|U$ is defined by putting $\tau(f)(V) = f(V) : E(V) = E^X(V) \to F(V) = (F|U)(V)$ for every $V \subset U$. Clearly τ is injective, so, to prove that τ is surjective, let $g : E \to F|U$ be a morphism in $\underline{S}(U)$. We then define $f : E^X \to F$ in $\underline{S}(X)$ by putting $f(V) = g(V) : E^X(V) \to (F|U)(V) = F(V)$ if $V \subset U$ and $f(V) : E^X(V) \to F(V)$ the trivial map if $V \not\subset U$ and hence $E^X(V) = 0$. Clearly $\tau(f) = g$, proving the assertion. We leave it to the reader to show that the bijection τ is functorial in E and F.

(2.27.) Corollary Any injective sheaf of abelian groups is flabby.

Proof For any $U \subset X$, let $c(U, Z)$ denote the constant sheaf on U with fibre Z. Let E be a sheaf of abelien groups on X, then we have

$$[c(U, Z)^X, E]_{\underline{S}(X)} = [c(U, Z), E|U]_{\underline{S}(U)} = [c_{\mathrm{Open}(U)}(Z), E|U]_{\underline{P}(U)} = E(U).$$

Now, if $V \subset U$ are open subsets of X, then there is an obvious monomorphism $c(V, Z)^X \subset c(U, Z)^X$, hence, if E is injective in $\underline{S}(X)$, this yields a surjective map $[c(U, Z)^X, E]_{\underline{S}(X)} = [c(V, Z)^X, E]_{\underline{S}(X)}$, i.e. a surjective map $E(U) \to E(V)$, which is easily seen to coincide with the restriction morphism E^U_V.

(2.28.) Lemma Let E be a sheaf of abelian groups, then $H^1(X, \gamma E) = 0$.

Proof Although one may prove quite easily that $H^1(U, \gamma E) = 0$ for <u>any</u> open covering U of X (by using some homotopy argument), we prefer to prove the result by an easy calculation, by showing that $H^1(U, \gamma E) = 0$ for an open covering $U = (U(x); x \in X)$, with $x \in U(x)$ for any x in X. In this case, let $\sigma \in C^1(U, \gamma E)$, then we may consider σ as a family

$$(s(u, v; x); (u, v) \in X \times X, x \in U(uv)),$$

since $\gamma E(U(uv)) = \prod\{E_x; x \in U(uv)\}$. If $\sigma \in \text{Ker}(d^1)$, then for all $u, v, w \in X$ and all x in $U(uvw)$, we have $s(v, w; x) - s(u, w; x) + s(u, v; x) = 0$. Now define $\tau = (s(u, v; x); u \in X, x \in U(u)) \in C^0(U, \gamma E)$, then $d^0(\tau) = (r(u, v; x); (u,v) \in X \times X, x \in U(uv))$ with $r(u, v; x) = s(x, v; x) - s(x, u; x) = s(u, v; x)$ - since $\sigma \in \text{Ker}(d^1)$. This proves that $\text{Im}(d^0) = \text{Ker}(d^1)$, i.e. $H^1(U, \gamma E) = 0$, indeed.

(2.29.) Lemma If E' is a flabby sheaf of abelian groups on X, then any exact sequence

$$0 \to E' \xrightarrow{f} E \xrightarrow{g} E'' \to 0$$

in $\underline{S}(X)$ is exact in $\underline{P}(X)$.

Proof We have to show that $g(U) : E(U) \to E''(U)$ is surjective for all open subsets U of X. We may assume $U = X$. Let G denote the set of all (U, s), where U is open in X and $s \in E(U)$, such that $g(U)(s) = s''|U$, where s'' is an arbitrary but fixed section in $E''(X)$. The set G may be inductively ordered in the obvious way, hence it possesses a maximal element (U, s). We want to show that $U = X$. So, let us assume $x \in X - U$ and pick a section $t \in E(V)$ over some open neighborhood V of x, such that $g(V)(t)$ $= s''|V$, then $g(U \cap V)(t|U \cap V) = g(U \cap V)(s|U \cap V)$, so $t|U \cap V - s|U \cap V \in$ $E'(U \cap V)$ (since E' is flabby). Let $s_1 = t - t'|V \in E(V)$, hence $s_1|U \cap V =$ $s|U \cap V$, i.e. we may find $s_2 \in E(U \cap V)$ such that $s_2|U = s$ and such that $g(U \cup V)(s_2) = s''|U \cup V$. This proves that (U, s) was not maximal, whence the assertion.

(2.30.) Corollary Let E be a flabby sheaf of abelian groups, then $H^1(X, E) = 0$.

Proof The exact sequence $0 \rightarrow E \rightarrow \gamma E \rightarrow E'' \rightarrow 0$ in $\underline{S}(X)$, with $E'' =$ $\mathrm{Coker}(E \rightarrow \gamma E)$ yields a cohomology sequence

$$0 \rightarrow E(X) \rightarrow \gamma E(X) \rightarrow E''(X) \rightarrow H^1(X, E) \rightarrow 0,$$

since $H^1(X, \gamma E) = 0$ by (2.28.). On the other hand, $\mathrm{Coker}(\gamma E(X) \rightarrow E''(X)) =$ 0 by (2.29.), since E is assumed to be flabby. This proves the assertion.

(2.31.) Corollary If E is an injective sheaf of abelian groups, then $H^1(X, E) = 0$.

Proof This is an immediate consequence of (2.30.) and (2.27.).

(2.32.) Let us conclude this section with a brief indication of some further structural results about the categories $\underline{P}(B)$ and $\underline{S}(B)$. We have already mentioned that both $\underline{P}(B)$ and $\underline{S}(B)$ are abelian categories. One may prove however, see Grothendieck's Tohoku paper [Gr] for example, that these are actually <u>Grothendieck categories</u> - see any (good) textbook on category theory for the definition and main features of this notion. In fact, we have already verified in (2.24.) that $\underline{S}(B)$ possesses enough injectives, and the analogous fact for presheaves is even easier to prove. The relationship between exactness properties in $\underline{P}(B)$ and $\underline{S}(B)$, indicated by the different behaviour of kernels and cokernels, is strongly linked to the fact that $\underline{S}(B)$ is a so-called <u>Giraud subcategory</u> of $\underline{P}(B)$. Let us recall briefly some of the machinery involved.

(2.33.) Let $\underline{P}$ be a complete Grothendieck category and $\underline{S}$ a full subcategory of $\underline{P}$ with canonical inclusion $i : \underline{S} \to \underline{P}$, then we say that $\underline{S}$ is a <u>reflective</u> subcategory of $\underline{P}$ if i possesses a left adjoint **a** (the "<u>reflector</u>" of $\underline{S}$ in $\underline{P}$), i.e. for every P in $\underline{P}$ and every S in $\underline{S}$ there should exist a functorial isomorphism $[P, iS]_{\underline{P}} \cong [aP, S]_{\underline{S}}$, with natural transformations p : $ai \to id_{\underline{S}}$ resp. q : $id_{\underline{P}} \to ia$. In particular, the couple $(aP, q_P : P \to iaP)$ is then universal in the following sense: for every $f : P \to iS$ in $\underline{P}$ with $S \in \underline{S}$, there is a unique $f' : iaP \to iS$ in $\underline{S}$ making the following diagram commutative:

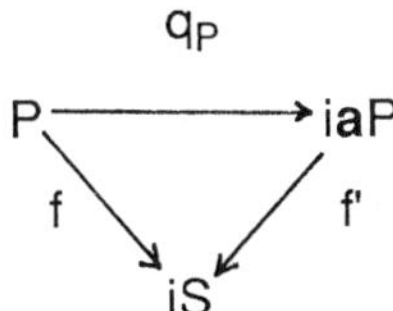

If the reflector **a** is left exact (hence exact, being a left adjoint to i), then we call $\underline{S}$ a <u>Giraud subcategory</u> of $\underline{P}$. Let **T** be the class of objects P in $\underline{P}$ such that $aC = 0$ and let **F** consist of the subobjects in $\underline{P}$ of objects in $\underline{S}$, then one may prove that the couple $(\mathbf{T}, \mathbf{F})$ determines a torsion theory in $\underline{P}$, such that its quotient category is equivalent to $\underline{S}$ (see [Ste, VV1] for example - we will come back to these notions later !). In fact, essentially <u>all</u> Giraud subcategories of $\underline{P}$ arise in this way.

(2.34.) As an example, it follows from (1.36.), (2.3.) and (2.16.) that $\underline{S}(B)$ is a Giraud subcategory of $\underline{P}(B)$ for any basis B of the topology on X, the reflector being the sheafification functor **a**. The class **F** consists exactly of the separated presheaves in $\underline{P}(B)$, due to (1.34.2.) and the obvious fact that a subpresheaf of a sheaf is separated. Actually, one may show that an object P in $\underline{P}$ (possessing a Giraud subcategory $\underline{S}$ and a reflector a) lies in F if and only if the reflector yields a monomorphism $q_P : P \to iaP$.

(2.35.) One may use the machinery of Giraud subcategories to derive some sheaf theoretic properties in an elegant, intrinsic way. For example, assume that $0 \to P' \to P \to P'' \to 0$ is an exact sequence in $\underline{P}$, then

(2.35.1.) if $P' \in \underline{S}$ and $P \in F$, then $P'' \in F$;

(2.35.2.) if $P'' \in F$ and $P \in \underline{S}$, then $P' \in \underline{S}$.

The second property implies for example that the kernel (in $\underline{P}(B)$) of a morphism in $\underline{S}(B)$ is a sheaf, and the first that the presheaf cokernel of a morphism $f : P' \to P$ where P' is a sheaf and P is separated is itself separated. In particular, if P is also a sheaf, then this yields that $\mathrm{Coker}_p(f) \subset \mathrm{Coker}_s(f)$. See also (2.5.).

3 SHEAVES OF MODULES

(3.1.) The main results in this text mostly deal with sheaves of modules over some subsets of certain schemes (see below). The notion of a sheaf of modules as well as the elementary operations that may be performed with them are introduced in the present Section, first in general and then adapted specifically to the algebro-geometric situation for which we need them. We will also study special types of sheaves of modules, a.o. quasicoherent, coherent, locally free and invertible sheaves. We will indicate briefly how these behave in the geometric context, but we refer to the E.G.A.'s or Hartshorne's book for more details and information. Some supplementary results will be presented as exercises at the end.

(3.2.) A (pre)sheaf of abelian groups A on some basis B of the topology on X is said to be a <u>(pre)sheaf of rings</u> on B, if A(U) is endowed with a ringstructure for each $U \in B$ and if for every $V \subset U$ in B the restriction morphism $A^U_V : A(U) \to A(V)$ is a ring homomorphism. If all of the A(U) are commutative, then we speak of a <u>(pre)sheaf of commutative rings</u> on B. As before, if B = Open(X), then we call A a (pre)sheaf of rings on X. If R is a commutative ring, then examples of sheaves of commutative rings on X are given by the constant sheaf with fibre R or the skyscraper sheaf $\sigma(y, R)$ with support y in X. If B is the canonical basis for the Zariski topology on Spec(R), then the structure presheaf $\underline{Q_R}$ is also a sheaf of commutative rings on B. Note also that the associated sheaf **a**A of a presheaf of rings is a sheaf of rings. Indeed, the sheaf space A~ is then a sheaf space of rings. Quite generally, we call a sheaf space of abelian groups (E, X, p) a <u>sheaf space of rings</u> if all of its stalks E_x $(x \in X)$ are rings and if the multiplication on these induce a continuous morphism $E \times_X E \to E$. Equivalently, if all the groups of sections $\Gamma(U, E)$ with U open in X are rings under the pointwise operations. The proof of this goes as that of (2.2.). In fact, one easily shows that there is a one-to-one correspondence between sheaves of rings and sheaf spaces of rings. If one defines morphisms between these objects in the natural way, this yields an equivalence between the corresponding categories.

(3.3.) Let A be a (pre)sheaf of rings on B and M a (pre)sheaf of abelian groups on B, then we call M a <u>(pre)sheaf of (left) A-modules</u> or sometimes simply a <u>(left) A-module</u>, when no ambiguity occurs, if the abelian group $M(U)$ is a (left) $A(U)$-module for each U in B and if for all $V \subset U$ in B the restriction $M^U{}_V : M(U) \to M(V)$ is $A^U{}_V$-semilinear, in the sense that for all a in $A(U)$ and all m in $M(U)$, we have $M^U{}_V(am) = A^U{}_V(a)M^U{}_V(m)$. It is clear that for each $x \in X$, we then have that M_x is a (left) A_x-module, due to the properties of inductive limits. It is also obvious what a <u>sheaf space of (left)-modules</u> over a sheaf space of rings should be and how to show that these correspond naturally to sheaves of modules over the corresponding sheaf of rings!

(3.4.) Morphisms of (pre)sheaves of A-modules on B are defined in the obvious way, i.e. we call a morphism $f : M \to N$ in $\underline{P}(B)$ or $\underline{S}(B)$ a <u>morphism of (pre)sheaves of A-modules</u>, if for each U in B the morphism $f(U) : M(U) \to N(U)$ is $A(U)$-linear. Since these maps may be composed in the obvious way, this yields new categories $\underline{P}(B, A)$ resp. $\underline{S}(B, A)$ consisting of presheaves of A-modules resp. its full subcategory consisting of sheaves, and corresponding maps. We reserve the notations $\underline{P}(X, A)$ and $\underline{S}(X, A)$ for the special case $B = \text{Open}(X)$. Let $\underline{C}$ denote one of these categories, then, taking stalks defines a functor

$$\tau_x : \underline{C} \to A_x\text{-mod},$$

for each x in X and taking sections over an open subset U a functor

$$\Gamma(U, -) : \underline{C} \to A(U)\text{-mod}.$$

We will use both notations $M(U)$ and $\Gamma(U, M)$ interchangeably, as before. One easily shows that the sheafification functor $\mathbf{a} : \underline{P}(B) \to \underline{S}(B)$ restricts to a functor $\mathbf{a} : \underline{P}(B, A) \to \underline{S}(B, \mathbf{a}A)$. In particular, it is thus fairly easy to see that if $f : M \to N$ is a morphism in $\underline{S}(B, A)$, then $\mathrm{Ker}(f)$, $\mathrm{Im}(f)$ and $\mathrm{Coker}(f)$, calculated in $\underline{S}(B)$, are actually lying in $\underline{S}(B, A)$, due to the fact that $\mathrm{Ker}(f)$, $\mathrm{Im}_p(f)$ and $\mathrm{Coker}_p(f)$ are trivially checked to belong to $\underline{P}(B, A)$ - we have chosen A to be a sheaf of rings, i.e. $\mathbf{a}A = A$, for simplicity's sake. From this one rather easily derives that $\underline{P}(B, A)$ and $\underline{S}(B, A)$ are abelian categories.
Actually, one may prove, cf. [Mi, Str, VV1]

(3.5.) Proposition The categories $\underline{P}(B, A)$ and $\underline{S}(B, A)$ are Grothendieck categories and $\underline{S}(B, A)$ is a Giraud subcategory of $\underline{P}(B, A)$.

For example, if $E \in \underline{S}(B, A)$, then one may embed E in an injective object of $\underline{S}(B, A)$ (an "injective A-module"), by applying the same method as in (2.24.). With the notations used there one just chooses I^x to be an injective A_x-module containing E_x, and one then constructs $\gamma((I^x; x \in X))$ as before. Note that for any family $(I^x; x \in X)$, where I^x lies in A_x-mod and any F in $\underline{S}(B, A)$, the map γ induces a bijection

$$[F, \gamma((I^x; x \in X))]_{\underline{S}(B, A)} = \Pi_x [F_x, I^x].$$

(3.6.) The most important examples of sheaves of rings or modules we will encounter in the sequel are (structure) sheaves over the spectrum of a commutative ring R. Recall that if R is a commutative ring and if B is the canonical basis for the Zariski topology on $\mathrm{Spec}(R)$, then we denote by $\underline{Q}_R$ the (structure) sheaf (of rings) on B. The machinery developed in the foregoing Sections allows us to define an associated sheaf $\underline{Q}_R$ on

Spec(R) (by first passing to the associated sheaf space of rings and by letting $\underline{O}_R$ be the image of $(\underline{Q}_R)^{\sim}$ by the functor $\Gamma : \underline{\mathbf{Sh}}(X) \to \underline{\mathbf{S}}(X)$). The sheaf of rings $\underline{O}_R$ is called the <u>structure sheaf</u> on Spec(R). Similarly, let M be an R-module, then we may define a presheaf $\underline{Q}_M$ on B by putting $\underline{Q}_M(X(s)) = M_s = M \otimes_R R_s$, the module of fractions of M at the multiplicative subset of R generated by s and with restriction morphisms as in (1.20.). We denote by $\underline{O}_M$ the associated sheaf on Spec(R). Since clearly $\underline{Q}_M$ is a presheaf of $\underline{Q}_R$-modules on B (since M_s is an R_s-module for each nonzero $s \in R$, this structure being compatible with the restriction morphisms), we find that $\underline{O}_M$ is a sheaf of $\underline{O}_R$-modules on Spec(R).

The main properties of $\underline{O}_M$ are given by:

(3.7.) Proposition Let M be an R-module, s a nonzero element of R and p a prime ideal of R, then

(3.7.1.) the sections of $\underline{O}_M$ over X(s) are given by $\Gamma(X(s), \underline{O}_M) = M_s$;

(3.7.2.) the global sections of $\underline{O}_M$ are given by $\Gamma(\mathrm{Spec}(R), \underline{O}_M) = M$;

(3.7.3.) the stalk of $\underline{O}_M$ in p is $\underline{O}_{M,p} = M_p$.

Proof The only non trivial statement is the first, since (2) follows from (1) with s = 1 and (3) follows immediately from $\underline{O}_{R,p} = \underline{Q}_{R,p} = R_p$, cf. (1.22.4.), since $\underline{Q}_M(X(s)) = \underline{Q}_R(X(s)) \otimes_R M$ for all basic open neighborhoods X(s) of p and since the tensorproduct commutes with inductive limits.

To prove (1), note that the functor **a** or Γ yields an obvious R-linear map $\phi_s : M_s = \underline{Q}_M(X(s)) \to \Gamma(X(s), \underline{O}_M)$, where we view $\underline{O}_M$ as a sheaf space, the map ϕ_s being determined by sending $t \in M_s$ to $t^{\sim} : X(s) \to \underline{O}_M$, given by $t^{\sim}(p) = t_p$, the image of t in M_p under the canonical map $j_p : M_s \to M_p$. We claim that the induced map $j = (j_p; p \in X(s)) : M_s \to \Pi_p M_p$ is injective, showing that ϕ_s is injective too. Indeed, pick a/s^n and b/s^m in M_s such that $j_p(a/s^n) = j_p(b/s^m)$ for all p in X(s), then for all such p there exists $t \notin p$ with $t(s^m a - s^n b) = 0$, hence $I = \mathrm{Ann}(s^m a - s^n b) \not\subset p$, i.e. $X(s) \subset X(I)$. But then we may find a positive integer u such that $s^u \in I$, hence $s^u(s^m a - s^n b) = 0$ and $a/s^n = b/s^m$, indeed.

To prove that ϕ_s is surjective, pick $\tau \in \Gamma(X(s), \underline{Q}_M)$, then, using the canonical basis for the Zariski topology, we find a family $(s(i) = s_i; i \in I)$ of elements of R such that $\{X(s_i); i \in I\}$ is an open covering of $X(s)$ and such that $\tau|X(s_i) = \phi_i(a_i/s_i)$ for some $a_i/s_i \in M_{s(i)}$, where we write $\phi_i = f_{s(i)}$ (note that $X(s_i) = X(s_i^n)$ for any n !). Moreover, since $X(s)$ is easily seen to be quasicompact, we may assume I to be finite (to prove this, note that if $X(s)$ $= \cup_i X(s_i) = X(\Sigma_i R s_i)$, then $\mathrm{rad}(Rs) = \mathrm{rad}(\Sigma_i R s_i)$, hence $s^n = \Sigma_i r_i s_i$ for some positive integer n, with $r_i = 0$ for all i outside a finite subset J of I, and we the obtain that the $X(s_i)$ with $i \in J$ also cover $X(s)$!). Of course a_i/s_i and a_j/s_j have the same image in $M_{s(i)s(j)}$ for each pair of indices i, j, and since there is only a finite number of these, we may find a single positive integer N such that $(s_i s_j)^N (s_j a_i - s_i a_j) = 0$, i.e. we find that $s_j^{N+1}(s_i^N a_i) = s_i^{N+1} s_j^N a_j)$ for all i, j. Write $s^n = \Sigma r_i s_i^{N+1}$ as above and put $a = \Sigma r_i s_i^N a_i \in M$, then $s^n(s_j^N a_j) = \Sigma r_i s_i^{N+1} s_j^N a_j = \Sigma r_i s_j^{N+1} s_i^N a_i = s_j^{N+1} a$. Hence $a/s^n = \tau'$ represents τ on each $X(s_i)$, since $a_j/s_j = s_j^N a_j/s_j^{N+1}$ in $M_{s(j)}$, so $\phi_s(\tau') = \tau$ and ϕ_s is surjective indeed.

Let us now calculate the sections of $\underline{Q}_M$ over an arbitrary open subset $X(I)$ of the Zariski topology on $\mathrm{Spec}(R)$. For any $M \in R\text{-mod}$, denote by $\sigma_I M$ the R-submodule of M consisting of all $m \in M$ such that $I^n m = 0$ for some positive integer n. If R is noetherian, we have $\mathrm{rad}(I)^N \subset I$ for some positive integer N, hence it is clear that for any R-module M we have $\sigma_I M = \sigma_{\mathrm{rad}(I)} M$.

(3.8.) Lemma Let R be a noetherian ring, I an ideal of R and M an R-module, then the canonical map $\lim \mathrm{Hom}_R(I^n, M) \to \lim \mathrm{Hom}_R(I^n, M/\sigma_I M)$ is an isomorphism.

Proof Denote by $p : M \to M' = M/\sigma_I M$ the natural quotient map, then the morphism $\phi : \lim \mathrm{Hom}_R(I^n, M) \to \lim \mathrm{Hom}_R(I^n, M')$ is determined by sending an R-linear map $f : I^n \to M$ to the composition $pf : I^n \to M'$. It is easy to see that ϕ is injective. Indeed, if $\phi(\alpha) = \phi(\beta)$ for some α resp. β represented by $f : I^n \to M$ resp. $g : I^m \to M$, then by definition we may find a positive integer N larger than m and n, such that $p(f|I^N) = p(g|I^N)$. Replace f and g by their restrictions to this I^N, then they still represent α and β, and for $h = f - g : I^N \to M$ we have $ph = 0$, i.e. $h(I^N) \to \sigma_I M$. Since I^N is finitely generated, there exists a positive integer N' such that $h(I^{N+N'}) = I^{N'}h(I^N) = 0$. Hence $h|I^{N+N'} = 0$ and $\alpha = \beta$, indeed.

The surjectivity of ϕ is a little more delicate. Since I is finitely generated, say by $x_1, \ldots , x_n$, there exists an R-linear surjective map $\pi : R^n \to I$, sending the canonical basis element e_i in R^n to x_i. Denote the kernel of p by K and let $f : I \to M'$ represent an element of $\lim \mathrm{Hom}_R(I^n, M')$, then there exists an R-linear morphism $g : R^n \to M$ extending f, i.e. such that the following diagram is commutative :

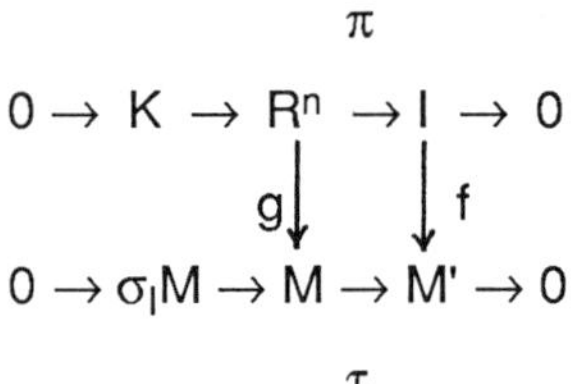

$$\begin{array}{ccccccccc}
 & & & & \pi & & & & \\
0 & \to & K & \to & R^n & \to & I & \to & 0 \\
 & & & & {\scriptstyle g}\downarrow & & \downarrow{\scriptstyle f} & & \\
0 & \to & \sigma_I M & \to & M & \to & M' & \to & 0 \\
 & & & & \tau & & & &
\end{array}$$

Indeed, just pick $m_i \in M$ such that $\tau(m_i) = f(x_i)$ and define g by $g(e_i) = m_i$. It follows that g maps K into $\sigma_I M$, hence $g(I^r K) = I^r g(K) = 0$ for some positive integer r, since $K \subset R^n$ is finitely generated. Since R is noetherian, we may apply Krull's Theorem (cf. [AM, 10.11.] for example), to obtain the existence of an $s \geq r$ such that $I^s R^n \cap K \subset I^r K$. Since $\mathrm{Ker}(\pi|I^s R^n) = I^s R^n \cap K \subset I^r K$, we have that $g_1 = g|I^s R^n$ factorizes through $\pi(I^s R^n) = I^{s+1} \subset I$, hence there exists an R-linear $f_1 : I^{s+1} \to M$ with the property that $f_1(p|I^s R^n) = g_1$, hence $\tau f_1(\pi|I^s R^n) = \tau g_1 = (f|I^{s+1})(\pi|I^s R^n)$, i.e. $\tau f_1 = f|I^{s+1}$, since π is surjective.

In other words, the class of the map f_1 in $\lim \mathrm{Hom}_R(I^n, M)$ maps onto the class of f in $\lim \mathrm{Hom}_R(I^n, M')$, so f is surjective as well.

We will be quite sketchy in the proof of the following result, since it will be considerably strengthened as well as be given a much easier proof (using some torsion-theoretic tricks) below. The reader is even invited to skip its proof at this moment.

(3.9.) Proposition (Deligne's formula) Let R be a noetherian ring and M an R-module, then for any ideal I in R we have

$$\Gamma(X(I), \underline{O}_M) = \lim \mathrm{Hom}_R(I^n, M).$$

Proof Since R is noetherian, I is finitely generated, say by some finite family $(s(i) = s_i;\ i = 1, \dots , k)$ of elements of R. Since the $X(s_i)$ cover $X(I)$, writing M_i for $M_{s(i)}$ and M_{ij} for $M_{s(i)s(j)}$, we get an exact sequence

$$0 \to \Gamma(X(I), \underline{O}_M) \xrightarrow{\alpha} \prod M_i \xrightarrow{\beta} \prod M_{ij} ,$$

with obvious maps.

There clearly is a map $\mu : \lim \mathrm{Hom}_R(I^n, M) \to \Gamma(X(I), \underline{O}_M)$ defined as follows. If the R-linear map $f : I^n \to M$ represents an element a in the first member, then for any index i, it yields by localizing an R-linear map $f_i : R_{s(i)} = (I^n)_{s(i)} \to M_{s(i)} = M_i$, and we map a to the family $(f_i(1))$ in $\mathrm{Ker}(\beta)$. One easily checks this to be well defined.

Conversely, let us define a map $v : \Gamma(X(I), \underline{O}_M) \to \lim \mathrm{Hom}_R(I^n, M)$ as follows. Choose $(\alpha_i = a_i/s_i^n) \in \Gamma(X(I), \underline{O}_M) = \mathrm{Ker}(\beta)$, where we may assume $n = 1$. Let $L_1 = R_{s(1)}$, then we may define an R-linear map $f_1 : L_1 \to M/\sigma_1 M$, by sending rs_1 in L_1 to the image of ra_1 in $M/\sigma_1 M$. By the previous Lemma, we may find a positive integer n and an R-linear $g_1 : Rs_1^n = L_1^n \to M$ making the following diagram commutative:

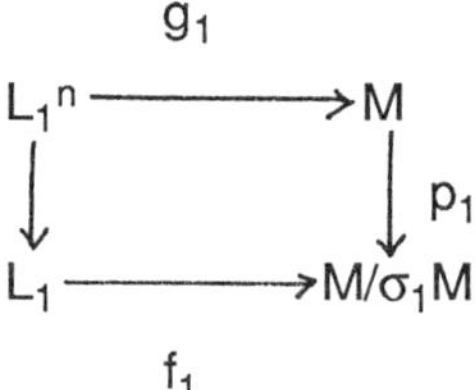

Let $b_1 = g_1(s_1^n)$, then, up to replacing a_1 by b_1 and s_1 by s_1^n, we thus obtain from g_1 an R-linear map $h_1 : Rs_1 \to M$ such that $\alpha_1 = a_1/s_1$ with $a_1 = h_1(s_1)$. Clearly, one easily checks that we may even assume that $a_i s_j = a_j s_i$ for all i, j.

Let us now argue by induction. Assume that for some $t < k$ we have already defined an R-linear map $h_t : L_t \to M$ and represented the α_i by a_i/s_i, with $s_i a_j = s_j a_i$ for each pair of indices i, j, such that $L_t = \Sigma_{i \leq t} Rs_i$ and $h_t(s_j) = a_i$ for each $i \leq t$, then we want to define $h_{t+1} : L_{t+1} \to M$ with similar properties. One does this by first constructing $f_{t+1} : Rs_{t+1} \to M/\sigma_{t+1}M$, where $\sigma_{t+1}M = \sigma_{Rs(t+1)}M$, exactly as we did for f_1, where we may have to change the s_i' s, getting a "new" L_t if necessary. We thus find representations a_i/s_i of the α_i with $s_i a_j = s_j a_i$ for each i, j together with $h_t : L_t \to M$ and $f_{t+1} : Rs_{t+1} \to M$ such that $h_t(s_i) = a_i$ for $i \leq t$ and $f_{t+1}(s_{t+1}) = a_{t+1}$. Let $L_{t+1} = L_t + Rs_{t+1}$ and denote by $p_{t+1} : M \to M/\sigma_{t+1}M$ the quotient map, then we define an R-linear $l_{t+1} : L_{t+1} \to M/\sigma_{t+1}M$ by $l_{t+1}|L_t = p_{t+1}h_t$ and $l_{t+1}|Rs_{t+1} = p_{t+1}f_{t+1}$; we leave it to the reader to check that this is well defined. Again applying the foregoing Lemma allows us to "lift" l_{t+1} to some $g_{t+1} : (L_{t+1})^n \to M$. Moreover, changing representatives if necessary, this last map may be modified to a map $h_{t+1} : (L_{t+1})^n \to M$, with the same nice properties as the previously constructed h_i.

By induction we thus find an R-linear map $h_k : L_k \to M$ such that L_k is generated by $s_1, \ldots, s_k$ with the property that $h_k(s_i)/s_i = \alpha_i$ for all i. Since the s_i generating L_k have been obtained by taking successive powers of the original s_i, clearly $rad(L_k) = rad(I)$, so $I^n \subset L_k$ for some positive integer

n. The map $v : \Gamma(X(I), \underline{O}_M) \to \lim \mathrm{Hom}_R(I^n, M)$ is now defined by sending the family $(\alpha_i; i = 1, \dots, k)$ to the class of $h_k|I^n : I^n \to M$ in $\lim \mathrm{Hom}_R(I^n, M)$. We leave it as an exercise to the reader to verify that μ and v are inverse to each other, proving the assertion.

(3.10.) From now on, assume throughout that $(X, \underline{O}_X)$ is a ringed space, i.e. X is some topological space and $\underline{O}_X$ a sheaf of (commutative) rings on X. The example to keep in mind is of course $(\mathrm{Spec}(R), \underline{O}_R)$, where R is a commutative ring. Of course, we could work more generally with a basis for the topology on X as well, but we leave it to the reader to verify the statements below in this case. Unless otherwise mentioned, all sheaves will be sheaves of $\underline{O}_X$-modules on X. If no ambiguity arises, for any pair of open subsets $V \subset U$ and any s in $E(U) = \Gamma(U, E)$ for some sheaf E, we will write $s|V$ for the restriction $E^U_V(s) \in E(V)$. Finally, we write $\mathrm{Hom}(E, F)$ or $[E, F]$ for the set of morphisms $E \to F$ in $\underline{S}(X, \underline{O}_X)$. Of course, $\mathrm{Hom}(E, F) \in \Gamma(X, \underline{O}_X)$-mod.

Most of the usual operations performed with modules over a ring may also be considered within $\underline{S}(X, \underline{O}_X)$. We have already mentioned the fact that $\underline{S}(X, \underline{O}_X)$ is a Grothendieck category. In particular, if $(E_i; i \in I)$ is a family of $\underline{O}_X$-modules, then we may construct presheaves in $\underline{P}(X, \underline{O}_X)$ by sending any open subset U of X to $\prod E_i(U)$ resp $\oplus E_i(U)$, with obvious restriction morphisms. The first construction yields a sheaf of $\underline{O}_X$-modules, denoted by $\prod_{i \in I} E_i$ and called the <u>product</u> or product sheaf of the E_i, the second yields a separated presheaf and we denote by $\oplus_{i \in I} E_i$ its associated sheaf, the <u>direct sum</u> or direct sum sheaf of the E_i. Of course, if I is finite, these notions coincide. Moreover, one may show (exercise!) that $\oplus$ and $\prod$ really behave as products resp. direct sums (coproducts) in $\underline{S}(X, \underline{O}_X)$, since we have

$$[\oplus_i E_i, F] = \prod_i [E_i, F]$$

resp.

$$[F, \prod_i E_i] = \prod_i [F, E_i].$$

Projective and inductive limits behave similarly. The obvious construction of a projective limit does not require sheafification, whereas that of inductive limits does. For details : see the literature.

(3.11.) Besides the operations $\oplus$ and $\prod$ which may be defined in any Grothendieck category, there are two more internal operations which play an important role in sheaf theory.

First, if E and F are in $\underline{S}(X, \underline{O}_X)$, define a presheaf on X by sending $U \in$ Open(X) to $E(U) \otimes F(U)$ (tensorproduct over $\underline{O}_X(U)$) and with obvious restriction morphisms. In general this is not a sheaf and we denote the associated sheaf of $\underline{O}_X$-modules by $E \otimes F$. We call it the <u>tensorproduct</u> of E and F (over $\underline{O}_X$). Clearly, $(E \otimes F)_x = E_x \otimes F_x$ for any $x \in X$.

On the other hand, we may also define a presheaf on X by sending $U \in$ Open(X) to $[E|U, F|U]$ (as $\underline{O}_X|U$-modules), again with obvious restriction maps. In this case this immediately yields a sheaf of $\underline{O}_X$-modules on X, which we denote by **Hom**(E, F). If $F = \underline{O}_X$, then we write $F^* = $ **Hom**$(E, \underline{O}_X)$ and we call it the <u>dual sheaf</u> of E.

Note that we do not necessarily have **Hom**$(E, F)_x = [E_x, F_x]$ for any x in X, so in particular we do not necessarily have $(E^*)_x = (E_x)^*$ (the latter dual in $\underline{O}_{X,x}$-mod!). See (3.17.) however...

The bifunctors $\otimes$ and **Hom** behave in many respects as their module counterparts, as far as exactness properties etc. are concerned. We leave details to the reader and again we refer to the literature.

(3.12.) Example Let M and N be R-modules for some ring R, then we claim that over $(\mathrm{Spec}(R), \underline{O}_R)$ there is an obvious isomorphism $\underline{O}_M \otimes \underline{O}_N =$

$\underline{O}_{M \otimes N}$. Indeed, $\underline{O}_M \otimes \underline{O}_N$ is the sheaf associated to the presheaf which sends the open subset $X(s) \subset \mathrm{Spec}(R)$ to $\underline{O}_M(X(s)) \otimes \underline{O}_N(X(s)) = M_s \otimes N_s = (M \otimes N)_s = \underline{O}_{M \otimes N}(X(s))$!

On the other hand, we do not necessarily have $\mathbf{Hom}(\underline{O}_M, \underline{O}_N) = \underline{O}_{[M, N]}$ as one might expect. However, this is true if M is finitely presented, essentially because associating $\underline{O}_M$ to M defines an exact fully faithful functor $\underline{O}_{(\,-\,)} : R\text{-mod} \to \underline{S}(\mathrm{Spec}(R), \underline{O}_R)$, as follows easily from (3.7.) and the definitions and from the left exactness of $[M, -\,]$ and $\mathbf{Hom}(\underline{O}_M, -\,)$.

(3.13.) Let us now study special types of sheaves over the ringed space $(X, \underline{O}_X)$. A sheaf of $\underline{O}_X$-modules E is said to be <u>free</u> (of finite rank) if it is isomorphic to a sheaf of $\underline{O}_X$-modules of the form $\underline{O}_X^{(I)}$ for some (finite) index set I, where $\underline{O}_X^{(I)} = \oplus_{i \in I} Q_i$, with $Q_i = \underline{O}_X$ for each i in I.

More generally, we say that E is <u>locally free</u> (of finite rank) if X may be covered by open subsets U such that for each of these, $E|U$ is a free sheaf of $\underline{O}_X|U$-modules (of finite rank), i.e. each $x \in X$ possesses an open neighborhood U together with an isomorphism of $\underline{O}_X|U$-modules $\underline{O}_X^{(I)}|U \cong E|U$. If for each U we may take the same I and if $|I| = n$, then we say that E is <u>locally free of rank n</u>.

(3.14.) Locally free sheaves of finite rank behave very nicely. Let us give an example. If E and F are sheaves of $\underline{O}_X$-modules, then there is an obvious map $E^* \otimes F \to \mathbf{Hom}(E, F)$, which is induced locally (on the presheaf level) by the module map $M^* \otimes N \to [M, N]$ which sends $f \otimes m \in M^* \otimes N$ to $<f, n> \in [M, N]$ defined by $m \to f(m)n$ for any $m \in M$. If E or F is locally free of finite rank, then one may easily verify this map to be bijective by reducing (locally) to the case $E = \underline{O}_X^n$ or $F = \underline{O}_X^m$, where the result is obvious.

By abuse of language, let us call E <u>invertible</u> if it is locally free of rank one. Clearly E^* is then also invertible. Note also that we then have $E^* \otimes$

$E = \underline{O}_X$ up to isomorphism. Indeed, by the foregoing remarks, it suffices to check that **End**$(E) =:$ **Hom**$(E, E) = \underline{O}_X$. But, there is an obvious map $\underline{O}_X \rightarrow$ **Hom**(E, E) (determined by the $\underline{O}_X$-module structure on E) and again reducing locally to the case $E = \underline{O}_X$, it follows that this map is an isomorphism. Finally, note that the tensorproduct of two invertible sheaves of $\underline{O}_X$-modules is invertible as well. It thus follows that the set of isomorphism classes [E] of invertible sheaves of $\underline{O}_X$-modules E forms a group Pic$(X, \underline{O}_X)$ or Pic(X), the so-called <u>Picard group</u> of $(X, \underline{O}_X)$. The composition law in Pic$(X, \underline{O}_X)$ is given by $[E] \cdot [E'] = [E \otimes E']$ and we have $[E]^{-1} = [E^*]$, the identity element being $[\underline{O}_X]$.

Denote by $\underline{O}_X{}^*$ (sorry for the ambiguity!) the sheaf of invertible elements of $\underline{O}_X$, i.e. $\Gamma(U, \underline{O}_X{}^*) = \Gamma(U, \underline{O}_X)^*$ for any $U \in$ Open(X), then we obtain the following cohomological description of Pic$(X, \underline{O}_X)$, whose proof is briefly sketched:

(3.15.) Proposition There is an isomorphism of abelian groups

$$\mathrm{Pic}(X, \underline{O}_X) = H^1(X, \underline{O}_X{}^*).$$

Proof If $U \in$ Open(X), then we claim that $\Gamma(U, \underline{O}_X{}^*) = \Gamma(U, \underline{O}_X)^*$ may be identified with the group of automorphisms of the sheaf of $\underline{O}_X|U$-modules $\underline{O}_X|U$. Indeed, this follows from the fact that the canonical map $\underline{O}_X \rightarrow$ **Hom**$(\underline{O}_X, \underline{O}_X)$ is an isomorphism, i.e. $\Gamma(U, \underline{O}_X) = \mathrm{End}(\underline{O}_X|U)$ and so $\Gamma(U, \underline{O}_X)^* = \mathrm{Aut}(\underline{O}_X|U)$.

Let E be locally free of rank 1, i.e. there exists a covering $U = (U_i; i \in I)$ and isomorphisms $f_i : E|U_i \rightarrow \underline{O}_X|U_i$, hence for all $(i,j) \in I \times I$ an automorphism s_{ij} of $\underline{O}_X|U_{ij}$ given by

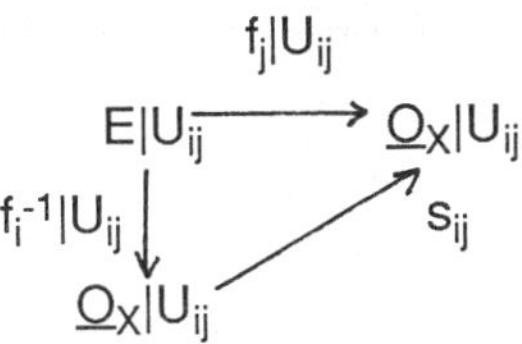

By the foregoing remark s_{ij} corresponds to some $t_{ij} \in \Gamma(U_{ij}, \underline{O}_X{}^*)$. An easy calculation shows that $s_{ij}s_{jk}=s_{ik}$ for any $i,j,k \in I$, hence $(t_{ij}; (i,j) \in I \times I) \in$ Ker(d^1) (notations as in (2.18.)!), i.e. we obtain an element in $H^1(U, \underline{O}_X{}^*)$. Conversely, if $(t_{ij}; (i,j) \in I \times I) \in$ Ker(d^1) or equivalently, the family of $s_{ij} : \underline{O}_X|U_{ij} \cong \underline{O}_X|U_{ij}$ is given, then we may associate to it a sheaf of $\underline{O}_X$-modules E by sending any $U \in$ Open(X) to the set $E(U) \subset \Pi\underline{O}_X(U \cap U_i)$ consisting of all families $(r_i; i \in I)$ with $s_{ij}(r_i) = r_j$ for all $i,j \in I$ (we have omitted the restriction maps) and with obvious restrictions. It is easy to see that E is a sheaf and that there are isomorphisms $g_i : E|U_i \to \underline{O}_X|U_i$ such that $f_j f_i^{-1} = s_{ij}$ on U_{ij}. If $(t_{ij}; (i,j) \in I \times I) \in$ Im(d^0), i.e. $t_{ij} = t_j t_i^{-1}$ for some $(t_i \in \Gamma(U_i, \underline{O}_X{}^*); i \in I)$, then the sheaf obtained from the $(s_{ij}t_{ij}; (i,j) \in I \times I)$ is isomorphic to E. So the two constructions may be seen to be inverse to each other, yielding a bijection $\phi_U : H^1(U, \underline{O}_X{}^*) \to \text{Pic}_U(X, \underline{O}_X)$, where $\text{Pic}_U(X, \underline{O}_X)$ is the subgroup of Pic(X, $\underline{O}_X$) consisting of all (isomorphism classes of) sheaves which are trivial on U, i.e. such that $E|U_i \cong \underline{O}_X|U_i$ for every $i \in I$. The maps ϕ_U are group homomorphisms and are easily checked to be compatible with refinement maps, i.e. the inclusions $\phi_U : H^1(U, \underline{O}_X{}^*) \to$ Pic(X, $\underline{O}_X$) yield an inclusion $\phi : H^1(X, \underline{O}_X{}^*) \subset$ Pic(X, $\underline{O}_X$) by passing to the inductive limit. On the other hand, since every invertible sheaf of $\underline{O}_X$-modules is trivial on some covering U, the map ϕ is also surjective, proving that ϕ is an isomorphism indeed.

(3.16.) Let us now study some other special types of sheaves over a ringed space (X, $\underline{O}_X$). A sheaf of $\underline{O}_X$-modules E is said to be <u>locally of finite type</u> resp. <u>locally finitely presented</u> if for each $x \in X$ there exists an open neighborhood U of x and an exact sequence of $\underline{O}_X|U$-modules of the form

$$\underline{O}_X{}^n|U \to E|U \to 0$$

for some positive integer n resp. of the form 

$$\underline{O}_X{}^p|U \to \underline{O}_X{}^q|U \to E|U \to 0$$

for some positive integers p and q.

The typical example to keep in mind is the case $(\mathrm{Spec}(R), \underline{O}_R)$ for some commutative ring R, where any finitely generated resp. finitely presented R-module M yields a sheaf of $\underline{O}_R$-modules $\underline{O}_M$, which is locally of finite type resp. finitely presented, due to the exactness of $M \to \underline{O}_M$ already mentioned in (3.12.).

(3.17.) Proposition Let E and F be sheaves of $\underline{O}_X$-modules on X and $x \in X$. If E is locally finitely presented, then there is an isomorphism

$$\mathbf{Hom}(E, F)_x \cong [E_x, F_x].$$

Proof For any $x \in X$, there is an open neighborhood U of x such that there exists an exact sequence of the form

$$\underline{O}_X{}^p|U \to \underline{O}_X{}^q|U \to E|U \to 0.$$

Applying $\mathbf{Hom}(-, F|U)$, which is left exact, and taking stalks at x yields an exact sequence of $\underline{O}_{X,x}$-modules

$$0 \to \mathbf{Hom}(E|U, F|U)_x \to \mathbf{Hom}(\underline{O}_X{}^q|U, F|U)_x \to \mathbf{Hom}(\underline{O}_X{}^p|U, F|U)_x.$$

On the other hand, first taking stalks at x and then applying the left exact functor $[\,-\,,\,F_x]$ yields a second exact sequence of $\underline{O}_{X,x}$-modules

$$0 \to [E_x, F_x] \to [(\underline{O}_{X,x})^q, F_x] \to [(\underline{O}_{X,x})^p, F_x].$$

These fit into an exact commutative diagram

$$0 \to \mathbf{Hom}(E|U, F|U)_x \to \mathbf{Hom}(\underline{O}_X{}^q|U, F|U)_x \to \mathbf{Hom}(\underline{O}_X{}^p|U, F|U)_x.$$

$$0 \longrightarrow [E_x, F_x] \longrightarrow [(\underline{O}_{X,x})^q, F_x] \longrightarrow [(\underline{O}_{X,x})^p, F_x].$$

the vertical morphisms denoting the isomorphisms $\mathbf{Hom}(\underline{O}_X{}^n|U, F|U)_x \cong (\mathbf{Hom}(\underline{O}_X|U, F|U)^n)_x \cong (F|U)^n{}_x \cong F_x{}^n \cong [(\underline{O}_{X,x})^n, F_x]$, hence $\mathbf{Hom}(E|U, F|U)_x = [E_x, F_x]$ (up to isomorphism). But $\mathbf{Hom}(E|U, F|U)_x = \mathbf{Hom}(E, F)_x$, as one easily sees, so this proves the assertion.

(3.18.) Corollary Let E be a locally finitely presented sheaf of $\underline{O}_X$-modules on X and suppose E_x is a free $\underline{O}_{X,x}$-module of rank n for some $x \in X$, then there exists an open neighborhood U of x such that $E|U$ is a free sheaf of $\underline{O}_X|U$-modules of rank n.

Proof Since E and of course $\underline{O}_X$ are both locally finitely presented, we have isomorphisms $\mathbf{Hom}(E, (\underline{O}_X)^n)_x \cong [E_x, (\underline{O}_{X,x})^n]$ resp. $\mathbf{Hom}((\underline{O}_X)^n, E)_x \cong [(\underline{O}_{X,x})^n, E_x]$. Since E_x is free of rank n, there are isomorphisms $u(x) : (\underline{O}_{X,x})^n \to E_x$ resp. $v(x) : E_x \to (\underline{O}_{X,x})^n$, which are mutually inverse to each other. On some common neighborhood V of x we may thus find $u : \underline{O}_X{}^n|V \to E|V$ resp. $v : E|V \to \underline{O}_X{}^n|V$ such that $u_x = u(x)$ rep. $v_x = v(x)$. But, since $(uv)_x = u(x)v(x) = \mathrm{id}$ and $(vu)_x = v(x)u(x) = \mathrm{id}$, the same argument yields an open neighborhood $U \subset V$ of x such that $uv|U = \mathrm{id}$ resp. $vu|U = \mathrm{id}$, i.e. $v|U : E|U \to \underline{O}_X{}^n|U = (\underline{O}_X|U)^n$ is an isomorphism.

(3.19.) The foregoing definitions and properties may give the impression that sheaves of $\underline{O}_X$-modules behave very much like modules over a ring. In many respects this is true, indeed. However, there are some fragrant differences. For example, in general, it is not true that for any sheaf of $\underline{O}_X$-modules E on X we may find index sets I, J together with an exact sequence of the form

$$\underline{O}_X^{(I)} \to \underline{O}_X^{(J)} \to E \to 0,$$

even if we restrict to any arbitrarily small open neighborhood of a point $x \in X$. If this condition holds, i.e. if every $x \in X$ possesses an open neighborhood U such that for some index sets I, J there exists an exact sequence of the form

$$\underline{O}_X^{(I)}|U \to \underline{O}_X^{(J)}|U \to E|U \to 0,$$

then we call E <u>quasicoherent</u>.

Obviously, every locally finitely presented sheaf of $\underline{O}_X$-modules is quasicoherent. It is also trivially verified that any commutative ring R yields quasicoherent sheaves of $\underline{O}_R$-modules $\underline{O}_M$ on Spec(R) for any R-module M - use a resolution $R^{(I)} \to R^{(J)} \to M \to 0$ of M and apply $\underline{O}_{(-)}$!

More precisely, we have the following result (cf. [GD], [Ha1], ...):

(3.20.) Theorem Over any commutative ring R, the following assertions are equivalent:

(3.20.1.) $E = \underline{O}_M$ for some R-module M (up to isomorphism);

(3.20.2.) E is a quasicoherent sheaf of $\underline{O}_R$-modules on Spec(R).

Proof One implication is trivial, so assume E to be quasicoherent. For any $s \in R$, the restriction map $M = \Gamma(\text{Spec}(R), E) \to \Gamma(X(s), E) = M(s)$ factorizes through $u(s) : M_s \to M(s)$, since s is invertible on M(s) and one easily checks these maps to glue together to a morphism $u : \underline{O}_M \to E$, which we want to prove isomorphic.

Let $m/s^p \in M_s$ be such that $u(s)(m/s^p) = 0$, then we want to show that $s^n m = 0$ for some positive integer n, for then $m/s^p = 0$ in M_s and $u(s)$ is injective. Since E is quasicoherent, we may cover $\mathrm{Spec}(R)$ by a finite (!) number of open subsets $X(i) = X(s(i))$ such that for each of these i there is an exact sequence

$$\underline{O}_R^{(I)}|X(i) \to \underline{O}_R^{(J)}|X(i) \to E|X(i) \to 0$$

for some index sets I and J (possibly depending on i). By taking sections over $X(i)$, we get a map

$$q(i) : F = R_{s(i)}^{(I)} \to R_{s(i)}^{(J)} = G$$

and we let $M(i) = \mathrm{Coker}(q(i))$. Since obviously $\underline{O}_F|X(i) = \underline{O}_R^{(I)}|X(i)$ and $\underline{O}_G|X(i) = \underline{O}_R^{(J)}|X(i)$, it follows that $E|X(i) = \underline{O}_{M(i)}|X(i)$ and $M(i) = M(s(i))$ resp. $M(ss(i)) = M(s(i))_s$, since we have $X(s) \cap X(i) = X(ss(i))$. So, if for some $m \in M$ we have $u(s)(m/1) = 0$, then $u_i(m|X(i)) = (m|X(i))/1 = 0$ in $M(i)_s$, where $u_i : M(i) \to M(i)_s$ is the canonical map, hence $s^{r(i)}(m|X(i)) = 0$. Choosing a single $r = r(i)$ for all i, we thus get that $(s^r m)|X(i) = 0$ for all i, hence $s^r m = 0$ as the $X(i)$ cover $X = \mathrm{Spec}(R)$. It follows that $m/1 = 0$ in M_s and from this one easily deduces the injectivity of $u(s)$.

Similarly, let $m \in \Gamma(X(s), E) = M(s)$, then we claim that there is a positive integer N such that $s^N m = n|X(s) = u(s)(n/1)$ for some $n \in M$, which will prove the surjectivity of $u(s)$, since then $u(s)(u/s^N) = m$.

Let $m_i = m|X(ss(i))$, then it follows that we may find some $n_i \in M(s)$, such that $n_i|X(ss(i)) = s^{q(i)} m_i$ for some positive integer $q(i)$. As a matter of fact, we may take the same $q = q(i)$ for all i. It follows that the restriction of $n_i|X(s(i)s(j)) - n_j(s(i)s(j))$ to $X(ss(i)s(j))$ vanishes, hence for some positive integer p it follows that $s^p n_i$ and $s^p n_j$ have the same restriction to $X(s(i)s(j))$

by the injectivity part of the proof. But then there exists $n \in \Gamma(\text{Spec}(R), E)$ = M with $n|X(i) = s^p n_i$ for all i. Let $N = p+q$, then for all i we get

$$s^N m|X(ss(i)) = s^p(s^q m|X(ss(i))) = s^p n_i|X(ss(i)) = n|X(ss(i)),$$

hence $s^N m = n|X(s)$ as the $X(s(i))$ cover $X(s)$. This finishes the proof.

(3.21.) Example Let $[E] \in \text{Pic}(\text{Spec}(R), \underline{O}_R)$, then E is locally free of rank 1 and so E is certainly quasicoherent, hence $E = \underline{O}_M$ for some essentially unique R-module M $(= \Gamma(\text{Spec}(R), E))$. Similarly, $E^* = \underline{O}_N$ for some R-module N. Now $E \otimes E^* = \underline{O}_R$ up to isomorphism, hence, since $\underline{O}_M \otimes \underline{O}_N = \underline{O}_{M \otimes N}$ by (3.12.), we find an isomorphism $M \otimes N = R$, i.e. M (and N) is an invertible (= locally projective of rank one) R-module. It thus easily follows that $\text{Pic}(\text{Spec}(R), \underline{O}_R)$ is isomorphic to the Picard group $\text{Pic}(R)$ of R.

(3.22.) Let us finish this Section with some remarks on coherent sheaves. A sheaf of $\underline{O}_X$-modules E on X is said to be <u>coherent</u> if it is locally of finite type and if for any subset U of X and any morphism of $\underline{O}_X|U$-modules $u : \underline{O}_X{}^n|U \to E|U$, the kernel Ker(u) is also locally of finite type. It follows immediately that any coherent sheaf is locally of finite presentation, the converse not necessarily being true in general.
One of the main technical Lemmas on coherent sheaves says that if

$$0 \to E' \to E \to E'' \to 0$$

is an exact sequence of sheaves of $\underline{O}_X$-modules on X, such that two of them are coherent, then so is the third. At this moment, the reader should know enough generalities on sheaves to handle this remark as an exercise (it reduces to some easy diagram-chasing, ...). In case of emergency, however, he is invited to consult [GD, (I.5.3.2.)] for example.

From this result one deduces immediately that the class of coherent sheaves of $\underline{O}_X$-modules on X is closed under taking finite direct sums, images, kernels and cokernels and the "fancy" operations $\otimes$ and **Hom**.
It is also an easy consequence of this that if $\underline{O}_X$ is coherent itself (as a sheaf of $\underline{O}_X$-modules), then E is coherent if and only if it is locally of finite presentation.

(3.23.) **Example** Let R be a noetherian ring and let $\underline{O}_M$ be a quasicoherent sheaf of $\underline{O}_R$-modules on Spec(R). It is clear that $\underline{O}_M$ is locally of finite type if M is finitely generated (or equivalently finitely presented, since R is noetherian). Conversely, if $\underline{O}_M$ is locally of finite type, then we may cover Spec(R) by a finite number of X(s(i)) such that $M_{s(i)}$ is a finitely generated $R_{s(i)}$-module. But then M is a finitely generated R-module itself, as one easily checks (cf. [Bo, (II.5.1. Prop 3)]).
We now claim that $\underline{O}_M$ is coherent if and only if M is finitely generated. Indeed, if $\underline{O}_M$ is coherent, then it is locally finitely presented, hence M is certainly finitely generated. For the converse, we essentially have to show that the kernel of any R-linear map $R^n \to M$ is finitely generated, but this follows from the fact theat R is noetherian.
We thus have also proved that $\underline{O}_M$ is a coherent sheaf (of $\underline{O}_R$-modules) on Spec(R), whenever R is noetherian.
In practice, it appears that coherent sheaves only work well, when some finiteness assumption (such as the noetherian hypothesis) is imposed on the base space. We will see examples of this below.

4 BASE CHANGE

(4.1.) In the previous Sections we have essentially always worked with sheaves and presheaves over a single topological space. In this Section, we will consider the behaviour of sheaves, if we pass from one topological space to another. The results described here prepare for Section 6, where we consider the possibility of extending (quasi) coherent sheaves on a generically closed subset of a scheme to the whole scheme. Results needed from algebraic geometry will only be briefly described. For more detailed information, we refer to the literature and in particular to Hartshorne's excellent text [Ha].

(4.2.) For simplicity's sake, we will work immediately with sheaves and presheaves of abelian groups. Let $f : X \to Y$ denote a continuous map between the topological spaces X and Y, then for any presheaf E on X, we may define a presheaf on Y by putting $(f_*E)(V) = E(f^{-1}(V))$ for any open subset V of Y and with the obvious restriction morphisms. This is easily seen to define a covariant functor $f_* : \underline{P}(X) \to \underline{P}(Y)$, which we call the <u>direct image</u> functor and which restricts to a functor $f_* : \underline{S}(X) \to \underline{S}(Y)$. Moreover, if $g : Y \to Z$ is another continuous map, then we have $(gf)_* = g_*f_*$.

(4.3.) If $x \in X$, then the morphisms $E(f^{-1}(U)) \to E_x$, where $f(x) \in U$ form an inductive system, which yields a morphism $(f_*E)_{f(x)} \to E_x$, which is in general neither injective nor surjective.

If f is injective however, then we may say more. Indeed, in this case, it is easily seen that f_*E induces E on X. In particular, $(f_*E)_{f(x)} = E_x$ for all $x \in X$. Note however that the support of f_*E, i.e. the set of all points $y \in Y$ where $(f_*E)_y \neq 0$ may in general be larger than that of E, since it may be the whole closure $\overline{X}$ of X in Y. If X is closed in Y (using f as an inclusion), then the supports of E and f_*E coincide. In particular, we then have $(f_*E)_y = 0$ for all $y \in Y - X$, i.e. f_*E is the extension of E by zero outside of Y.

If we choose some $x_0 \in X$ and we denote by $f : \overline{\{x_0\}} \to Y$ the inclusion of the closure of x_0 in Y, then for any abelian group G we recover the skyscraper sheaf $\sigma(x_0, G)$ defined in (1.33.) as a special case of the foregoing construction. Indeed, denote by $c(\overline{\{x_0\}}, G)$ the constant sheaf with fibre G on $\overline{\{x_0\}}$, cf. (1.32.), then one easily checks that $\sigma(x_0, G) = f_*c(\overline{\{x_0\}}, G)$.

(4.4.) Now, let F be a presheaf on Y, where we still assume $f : X \to Y$ to be a continuous map, then we may define a presheaf $f_p^{-1}F$ on X by putting

$$(f_p^{-1}F)(U) = \lim_{f(U) \subset V} F(V)$$

for any open subset U of X, the inductive limit being taken over all open subsets V of Y containing $f(U)$ and with obvious restriction morphisms. Since $f_p^{-1}F$ is not a sheaf in general, even if F is , we let $f^{-1}F$ (or $f_s^{-1}F$, whenever ambiguity arises) denote the associated sheaf $\mathbf{a}f_p^{-1}F$.

It is clear that $(f^{-1}F)(U)$ may be viewed as the collection of all families $(s_x; x \in U)$, where $s_x \in F_{f(x)}$ for each $x \in X$, with the property that there exists $V \in V_Y(f(x))$ and $W \in V_X(x)$ with $W \subset f^{-1}(V) \cap U$ such that for all $z \in W$ we have $s^z = s_{f(z)} \in F_{f(z)}$.

It follows easily from this that we thus obtain a covariant functor $f^{-1} : \underline{P}(Y) \to \underline{P}(X)$, which maps a morphism $f : F_1 \to F_2$ of presheaves on Y to the morphism $f^{-1}f : f^{-1}F \to f^{-1}F_2$ defined on any open subset U of X by sending $(s^x; x \in U) \in (f^{-1}F_1)(U)$ to $(f_x(s^x); x \in U) \in (f^{-1}F_2)(U)$. Of course, f also induces directly a functor $f_p^{-1} : \underline{P}(Y) \to \underline{P}(X)$ such that $f^{-1} = \mathbf{a}f_p^{-1}$. Given in the above form, however, it follows immediately that the identity on X induces the functor $\mathbf{a} : \underline{P}(X) \to \underline{S}(X)$ i.e. $\mathbf{a} = \mathrm{id}_X^{-1}$.

More generally, if $f : X \to Y$ is an inclusion, then we will usually write $F|X$ instead of $f^{-1}F$. In this case we have $(F|X)_x = F_x$ for all $x \in X$. In particular, if X is an open subset of Y, then the restriction $F|X$ thus defined coincides with that defined in Section 1. Finally note that the inverse image functor has the following property : if $g : Y \to Z$ is another continuous map, then $(gf)^{-1} = f^{-1}g^{-1}$.

(4.5.) The functors $f_* : \underline{S}(X) \to \underline{S}(Y)$ and $f^{-1} : \underline{S}(Y) \to \underline{S}(X)$ are adjoint to each other. Indeed, let E be a sheaf on X resp. F a sheaf on Y, then we define a morphism $[F, f_*E]_{\underline{S}(Y)} \to [f^{-1}F, E]_{\underline{S}(X)}$ as follows. Let $g : F \to f_*E$ be a morphism in $\underline{S}(Y)$ and let U be open in X, then for any open subset V of Y with $f(U) \subset V$, we have $U \subset f^{-1}(V)$, hence we get a map $F(V) \to (f_*E)(V) = E(f^{-1}(V)) \to E(U)$ and by passing to the inductive limit we obviously get a map $(f_p^{-1}F)(U) = \lim F(V) \to E(U)$. These maps glue together to a morphism $f_p^{-1}F \to E$, hence yield a morphism $f^{-1}F \to E$ by passing to the associated sheaf.

Conversely, let $h : f^{-1}F \to E$ be a morphism in $\underline{S}(X)$ and let V be an open subset of Y, then, denoting $f^{-1}(V) \subset X$ by U, we get a map $F(V) \to \lim F(W) \to (f^{-1}F)(U) \to E(U)$, where the limit is taken over all open subsets W of Y containing U. These maps glue together to a morphism $F \to f_*E$ in $\underline{S}(Y)$.

We leave it to the reader to verify that these procedures are inverse to each other and yield a functorial bijection $[F, f_*E]_{\underline{S}(Y)} \to [f^{-1}F, E]_{\underline{S}(X)}$ indeed.

(4.6.) If we work with sheaves of modules, then things work somewhat differently. Indeed, assume that X and Y are ringed spaces, i.e. the topological spaces X and Y are endowed with sheaves of rings $\underline{O}_X$ and $\underline{O}_Y$, then it may be verified that even for the notion of a morphism between such spaces defined below, there is no reason why $f^{-1}E$ should be a sheaf of $\underline{O}_X$-modules if E is a sheaf of $\underline{O}_Y$-modules. However, for such morphisms, f_*E is a sheaf of $\underline{O}_Y$-modules if E is a sheaf of $\underline{O}_X$-modules. Let us be a little more precise about this in the next paragraphs.

(4.7.) By definition, <u>a morphism of ringed spaces</u> $(X, \underline{O}_X) \to (Y, \underline{O}_Y)$ is a couple (f, θ), where $f : X \to Y$ is a continuous map and $\theta : \underline{O}_Y \to f_*\underline{O}_X$ is a morphism of sheaves of rings. These morphisms may be composed in the obvious way, thus yielding a category, the category of ringed spaces. It is readily verified that an isomorphism in this category is a pair (f, θ) as above, where now f is a homeomorphism and θ an isomorphism of sheaves of rings. For each $x \in X$ and $y = f(x)$, the map θ induces a morphism $\underline{O}_{Y,y} = \lim \underline{O}_Y(V) \to \lim \underline{O}_X(f^{-1}(V)) \to \underline{O}_{X,x}$ between the corresponding stalks. If for all $x \in X$ the stalk $\underline{O}_{X,x}$ is a local ring, then we call $(X, \underline{O}_X)$ a <u>locally ringed space</u> and if $(Y, \underline{O}_Y)$ is also a locally ringed space, then (f, θ) is a <u>morphism of locally ringed spaces</u> if the map $\underline{O}_{Y,y} \to \underline{O}_{X,x}$ is local for each $x \in X$, in the sense that the inverse image of the maximal ideal of $\underline{O}_{X,x}$ is the maximal ideal of $\underline{O}_{Y,y}$. This clearly defines the category of locally ringed spaces.

(4.8.) <u>The</u> example of a locally ringed space is of course the ringed space $(Spec(R), \underline{O}_R)$ associated to a commutative ring R. As we have pointed out before, there is a bijective correspondence between commutative rings and ringed spaces of this form, since we know that $\Gamma(Spec(R), \underline{O}_R) = R$. This correspondence is even functorial in the sense that the morphism sets $[R, S]$ and $[(Spec(S), \underline{O}_S), (Spec(R), \underline{O}_R)]$ may be identified for any pair of commutative rings R and S, where morphisms are taken in the category of rings resp. of locally ringed spaces.

Indeed, let $u : R \to S$ be a ring morphism, then it induces a map $^au :$ $Spec(S) \to Spec(R) : p \to u^{-1}(p)$, which is continuous, since for every ideal I of R we have $(^au)^{-1}(X_R(I)) = X_S(Su(I))$. We may define a map $u^{\sim} :$ $\underline{O}_R \to (^au)_*\underline{O}_S$ by putting for each $s \in R$

$$u^\sim(X_R(s)) : \underline{O}_R(X_R(s)) = R_s \to S_{u(s)} = \underline{O}_S(X_S(u(s))) = (^au)_*\underline{O}_S(X(s))$$

and then by passing to the whole sheaf, since $(^au)_*$ is easily checked to commute with the associated sheaf functor **a**. For any $q \in \mathrm{Spec}(S)$ and $p = (^au)(q) = u^{-1}(q)$, the stalk map $\underline{O}_{R,p} \to \underline{O}_{S,q}$ is just the ring morphism $R_p \to S_q$ induced by u, which is obviously local. This shows that any $u : R \to S$ yields a map of locally ringed spaces $(^au, u^\sim) : (\mathrm{Spec}(S), \underline{O}_S) \to (\mathrm{Spec}(R), \underline{O}_R)$.

Conversely, let $(f, \theta) : (\mathrm{Spec}(S), \underline{O}_S) \to (\mathrm{Spec}(R), \underline{O}_R)$ be a morphism of locally ringed spaces, then, taking global sections, θ induces a map $u = \theta(\mathrm{Spec}(R)) : R = \Gamma(\mathrm{Spec}(R), \underline{O}_R) \to \Gamma(\mathrm{Spec}(S), f_*\underline{O}_S) = S$ and one easily verifies that $(^au, u^\sim) = (f, \theta)$. Indeed, denote for any $q \in \mathrm{Spec}(S)$ and $p = f(q)$, by $\theta_q : R_p \to S_q$ the induced local ring homomorphism, then we have a commutative diagram

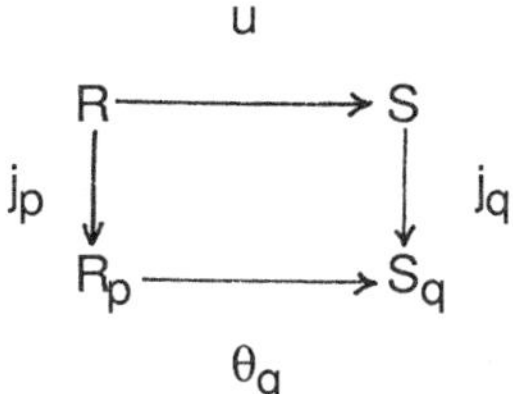

and since θ_q is local, we then obtain that $f(q) = p = u^{-1}(q)$, i.e. $^au = f$. The rest of the proof is left as an exercise to the reader.

(4.9.) More generally, define $(X, \underline{O}_X)$ to be an <u>affine scheme</u> if it is isomorphic (as a locally ringed space) to some $(\mathrm{Spec}(R), \underline{O}_R)$ for a commutative ring R. For example, if $s \in R$, then $(X(s), \underline{O}_R|X(s))$ is an affine scheme - as a matter of fact, it is isomorphic to the space $(\mathrm{Spec}(S), \underline{O}_S)$, where $S = R_s$, the localization of R at the multiplicative set generated by s.

A locally ringed space $(X, \underline{O}_X)$ is then called a <u>scheme</u>, if it is locally affine, i.e. if we may find an open covering $\{U_i ; i \in I\}$ of X, such that $(U_i, \underline{O}_X | U_i)$ is an affine scheme for all $i \in I$.

A <u>morphism of schemes</u> is just a morphism between the underlying locally ringed spaces and since these may be composed in the obvious way, this yields the so-called category of schemes. The functors Spec and Γ are adjoint for this category, in the sense that for any commutative ring R and any scheme $(X, \underline{O}_X)$ (actually for any locally ringed space $(X, \underline{O}_X)$), there is a bijective correspondence between the homomorphisms $[(X, \underline{O}_X), (\mathrm{Spec}(R), \underline{O}_R)]$ and $[R, \Gamma(X, \underline{O}_X)]$. The proof of this follows from the remarks made in (4.8.).

From now on, we will make the following conventions. If $(X, \underline{O}_X)$ is a scheme, then we will just say that X is a scheme, when no ambiguity arises. If $(f, \theta) : (X, \underline{O}_X) \to (Y, \underline{O}_Y)$ is a morphism of schemes, then we say that $f : X \to Y$ is a scheme morphism and we put $f^\# = \theta : \underline{O}_X \to f_* \underline{O}_X$.

(4.10.) Let us now look at base change with respect to morphisms of schemes. Assume that $f : X \to Y$ is a morphism of schemes (or, more generally, of ringed spaces!) and let E be an $\underline{O}_X$-module, then clearly $f_* E$ is an $f_* \underline{O}_X$-module, hence an $\underline{O}_Y$-module, through the morphism $f^\# : \underline{O}_Y \to f_* \underline{O}_X$. This defines a functor $f_* : \underline{S}(X, \underline{O}_X) \to \underline{S}(Y, \underline{O}_Y)$, which is easily seen to be left exact.

On the other hand, let F be an $\underline{O}_Y$-module, then $f^{-1}F$ is an $f^{-1}\underline{O}_X$-module. We have seen in (4.5.) that f_* and f^{-1} are adjoint to each other, hence the morphism $f^\# \in [\underline{O}_Y, f_* \underline{O}_X]$ corresponds to a morphism in $[f^{-1}\underline{O}_Y, \underline{O}_X]$, which is also a morphism of sheaves of rings (on X). We put $f^*F = f^{-1}F \otimes \underline{O}_X$, where the tensorproduct is over $f^{-1}\underline{O}_Y$, hence f^*F is naturally endowed with an $\underline{O}_X$-module structure. We leave it to the reader to check that this defines an exact functor $f^* : \underline{S}(Y, \underline{O}_Y) \to \underline{S}(X, \underline{O}_X)$, which commutes with tensorproducts and direct sums.

In particular, working in the general case for a moment, if $f : X \to Y$ is the inclusion of X into Y, and if the ringed space structure on X is induced by that of Y, i.e. $\underline{O}_X = \underline{O}_Y|X = f^{-1}\underline{O}_Y$, then the functors f^* and f^{-1} may be identified. In particular, we then have $E \otimes F|X = E|X \otimes F|X$ and $(\oplus_{i \in I}E_i)|X = \oplus_{i \in I}E_i|X$, for any pair of $\underline{O}_Y$-modules E, F and any family of $\underline{O}_Y$-modules $(E_i; i \in I)$.

Using the adjointness of f_* and f^{-1} and the universal properties of the tensorproduct, it is now an easy exercise to prove that the functors f^* and f_* are adjoint to each other, in the sense that for each $E \in \underline{S}(X, \underline{O}_X)$ resp. $F \in \underline{S}(Y, \underline{O}_Y)$, there is a functorial bijection

$$[f^*F, E] \cong [F, f_*E],$$

where the first morphism set is in $\underline{S}(X, \underline{O}_X)$ and the second in $\underline{S}(Y, \underline{O}_Y)$.

(4.11.) Example Let $u : R \to S$ be a morphism of commutative rings and let $f = {}^au : \text{Spec}(S) \to \text{Spec}(R)$ denote the associated scheme morphism. If M is an R-module and N an S-module, then $f_*\underline{O}_N = \underline{O}_{N'}$, where $N' = {}_RN$, i.e. N viewed as an R-module, and $f^*\underline{O}_M = \underline{O}_{M \otimes S}$.

(4.12.) The inverse image functor $f^* : \underline{S}(Y, \underline{O}_Y) \to \underline{S}(X, \underline{O}_X)$ associated to an arbitrary morphism $f : X \to Y$ of ringed spaces behaves well with respect to quasicoherent and coherent sheaves of modules.

Indeed, let F be quasicoherent on Y, then we claim that f^*F is quasicoherent on X. For if $x \in X$, then we may find $V \in V_Y(f(x))$ such that there is an exact sequence of the form

$$(*) \qquad \underline{O}_Y^{(I)}|V \to \underline{O}_Y^{(J)}|V \to F|V \to 0$$

for certain index sets I, J. Let $U = f^{-1}(V)$ and denote by $g : U \to V$ the restriction of f to U, then $(*)$ yields an exact sequence of $\underline{O}_X|U$-modules

$$(\text{**}) \qquad g^*(\underline{O}_Y{}^{(I)}|V) \to g^*(\underline{O}_Y{}^{(J)}|V) \to g^*(F|V) \to 0$$

But, since $jg = fi$, where $i : U \to X$ resp. $j : V \to Y$ denote the canonical inclusions, we find for all F that $g^*(F|V) = g^*j^*F = i^*f^*F = f^*F|U$. In particular, $g^*(\underline{O}_Y{}^{(I)}|V) = f^*(\underline{O}_Y{}^{(I)})|U = (f^*\underline{O}_Y)^{(I)}|U = \underline{O}_X{}^{(I)}|U$, hence (**) reduces to

$$\underline{O}_X{}^{(I)}|U \to \underline{O}_X{}^{(J)}|U \to f^*F|U \to 0,$$

proving that f^*F is quasicoherent, indeed.

On the other hand, if $\underline{O}_X$ is a coherent sheaf of rings and F is coherent, then the same argument, but with finite index sets I, J, yields that f^*F is a coherent sheaf of $\underline{O}_Y$-modules!

Note also that f^* maps locally free sheaves to locally free sheaves and in particular invertible sheaves to invertible sheaves. Since for any pair of $\underline{O}_X$-modules E and F we have $f^*(E \otimes F) = f^*E \otimes f^*F$, it follows that f^* induces a group homomorphism

$$\text{Pic}(Y, \underline{O}_Y) \to \text{Pic}(X, \underline{O}_X).$$

(4.13.) Example An open covering $\{U_i; i \in I\}$ of a scheme X is said to be _affine_ if $(U_i, \underline{O}_X|U_i)$ is an affine scheme for all $i \in I$. We call X _locally noetherian_ if it possesses an affine open covering $\{U_i; i \in I\}$ with $\Gamma(U_i, \underline{O}_X)$ noetherian for all $i \in I$. If such a covering exists with a finite index set I, then we call X _noetherian_. It is easy to see that X is noetherian if and only if it is locally noetherian and quasicompact. If X is locally noetherian, then $\underline{O}_X$ is a coherent sheaf of rings. Indeed, for each U in an open covering as above, the ring $R = \Gamma(U, \underline{O}_X)$ is noetherian and $\underline{O}_X|U = \underline{O}_U = \underline{O}_R$, which is coherent. Since coherence is a local property, this proves the assertion.

If $f : X \to Y$ is a morphism of schemes (with X and Y locally noetherian), then the remarks made in (4.12.) show that for any quasicoherent (resp. coherent) sheaf F on Y the sheaf f^*F on X is quasicoherent (resp. coherent) too. In this situation, this also follows from the fact that we may assume X and Y affine (arguing locally) and apply the explicit description of f^* in this case, cf. (4.11.).

(4.14.) Even when X and Y are schemes, it is easy to give examples of morphisms $f : X \to Y$ and a quasicoherent sheaf E on X such that f_*E is not quasicoherent. Several technical (but harmless) conditions on f, X and Y, guaranteeing that f_*E is coherent whenever E is, have been described in the literature, cf. [Ha], [GD], ... We also refer to the Exercises for results in this direction. Here, we will only look at the situation where f is the inclusion of an open subset X into Y. The more general case, where X is a generically closed subset of Y will be dealt with in Section 6, when the necessary background has been developed.

(4.15.) Lemma Let I be a finitely generated ideal of a commutative ring R and denote by $i : X(I) \subset \mathrm{Spec}(R)$ the canonical inclusion, then for any quasicoherent sheaf of $\underline{O}_R|X(I)$-modules on X(I), the direct image i_*E is quasicoherent on Spec(R).

Proof Since I is finitely generated, there exists an open covering $\{X(f_\alpha); 1 \leq \alpha \leq n\}$ of X(I). For any $1 \leq \alpha, \beta \leq n$, let $E_\alpha = i_*(E|X(f_\alpha))$ resp. $E_{\alpha\beta} = i_*(E|X(f_\alpha f_\beta))$, where we still denote by i the inclusion of $X(f_\alpha)$ or $X(f_\alpha f_\beta)$ in Spec(R). Since $X(f_\alpha)$ and $X(f_\alpha f_\beta)$ are affine, cf. (4.9.), it follows from (4.11.) that E_α and $E_{\alpha\beta}$ are quasicoherent sheaves on Spec(R).

For each open subset U of X, we have $E_\alpha(U) = E(X(f_\alpha) \cap U)$ and $E_{\alpha\beta}(U) = E(X(f_\alpha f_\beta) \cap U)$, hence we may define a morphism

$$d(U) : \oplus_\alpha E_\alpha(U) \to \oplus_{\alpha,\beta} E_{\alpha\beta}(U),$$

which sends $(s_\alpha; 1 \leq \alpha \leq n) \in \oplus_\alpha E(X(f_\alpha) \cap U)$ to $(s_{\alpha\beta}; 1 \leq \alpha, \beta \leq n) \in \oplus_{\alpha,\beta} E(X(f_\alpha f_\beta) \cap U)$ given by $s_{\alpha\beta} = s_\alpha | X(f_\alpha f_\beta) \cap U - s_\beta | X(f_\alpha f_\beta) \cap U$.

Since E is a sheaf, clearly $\mathrm{Ker}(d(U)) = E(X(I) \cap U) = i_* E(U)$. The $d(U)$ glue together to a presheaf morphism d and passing to the associated sheaves, this yields an exact sequence of $\underline{O}_R$-modules

$$(\text{*}) \qquad 0 \to i_* E \to \oplus_\alpha E_\alpha \to \oplus_{\alpha,\beta} E_{\alpha\beta},$$

hence $i_* E$ is quasicoherent.

(4.16.) Corollary Let U be a locally finite open subset of a scheme X, i.e. X may be covered by open affines $U_\alpha = \mathrm{Spec}(R_\alpha)$ such that $U \cap U_\alpha = X(I_\alpha) \subset \mathrm{Spec}(R_\alpha)$ for some finitely generated ideal I_α of R_α, and denote by $i : U \to X$ the canonical inclusion. If E is a quasicoherent sheaf of $\underline{O}_X | U$-modules on U, then $i_* E$ is quasicoherent.

Proof Choose an affine open covering $\{U_\alpha = \mathrm{Spec}(R_\alpha); \alpha \in I\}$ of X such that $V_\alpha = U_\alpha \cap U$ is of the form $X(I_\alpha)$ for some finitely generated ideal I_α of R_α, and denote by $i_\alpha : V_\alpha \to U_\alpha$ the canonical inclusion. Obviously, for each $\alpha \in I$ we have $(i_* E)|U_\alpha = i_{\alpha,*}(E|V_\alpha)$, hence $i_* E$ is quasicoherent, due to the fact that this property is local on X and that $i_{\alpha,*}(E|V_\alpha)$ is quasicoherent on U_α by the foregoing Lemma.

(4.17.) Notes

(4.17.1.) The foregoing result implies that for any locally noetherian scheme X and any open subset U of X with inclusion $i : U \to X$, the direct image $i_* E$ of a quasicoherent sheaf of $\underline{O}_X | U$-modules on U is quasicoherent on X, since the hypotheses are trivially satisfied.

(4.17.2.) Note also that for the extension $i_* E$ on X we have $i_* E|U = E$, as one trivially verifies. It follows (still for a locally finite subset U of the scheme X) that any quasicoherent sheaf on U extends to a quasicoherent sheaf on X. In Section 6 we will be concerned with classifying all of these extensions.

5 GENERICALLY CLOSED SETS

(5.1.) In this Section and the next ones, we will be concerned with the behaviour of (quasi) coherent sheaves on certain subsets of an arbitrary scheme. We have already indicated in (3.9.) how to calculate, for any module M over a (commutative) noetherian ring R, the sections of the sheaf $\underline{O}_M$ over some open subset X(I) of Spec(R). For our purposes however, this result is far from satisfactory. Indeed, first of all it requires the ground ring R to be noetherian, which is sometimes a rather severe restriction, e.g. if we want to consider Krull domains or Krull schemes. In [VV2] however, F. Van Oystaeyen and the author only dealt with the noetherian situation, which made the results there somewhat less suitable for applications in the theory of relative invariants, at least in the non-noetherian situation. As a matter of fact, Krull domains etc. could not be treated there. On the other hand, we will frequently have to work with subsets of Spec(R) or an arbitrary scheme, which are not open. Again, take a Krull domain R, then the study of reflexive R-modules is tightly connected to the subset $Spec^{(1)}(R) \subset Spec(R)$ consisting of all primes p of R with $ht(p) \leq 1$, cf. [Au, Or1, Yu]. This leads us to introduce generically closed subsets of schemes and to study finiteness conditions with respect to these.

(5.2.) A subset Y of a topological space X is said to be <u>generically closed</u> or <u>closed under generization</u> if for any $x \in X$, we have $x \in Y$ if some $y \in Y$ lies in the Zariski closure of $\{x\}$. Note that if X = Spec(R) for some ring R, then $Y \subset X$ is generically closed if and only if for any $p \in X$ with $p \subset q$ for some $q \in Y$, we have $p \in Y$ as well. If X is a scheme and Y a subset of X, then clearly Y is generically closed in X if and only if $Y \cap U$ is generically closed in U for any open (affine) subset U of X. This follows from the (easily verified) facts that any open subset of a topological space

X is generically closed and the fact that the set of all generically closed subsets of X is closed under intersections.

(5.3.) Any subset Y of a topological space X is contained in a unique generically closed subset $Y^\wedge$ of X, minimal with this property. Indeed, $Y^\wedge$ is just the intersection of all generically closed subsets of X containing Y. Equivalently, $Y^\wedge$ consists of all x in X such that y is contained in the closure of $\{x\}$ for some y in Y.

The operator $\wedge$ has the following properties:

(5.3.1.) if $Y \subset X$ is open (or more generally, generically closed), then $Y = Y^\wedge$;

(5.3.2.) if $Y \subset Z$, then $Y^\wedge \subset Z^\wedge$;

(5.3.3.) $(\cap_i Y_i)^\wedge \subset \cap_i Y_i^\wedge$;

(5.3.4.) $(\cup_i Y_i)^\wedge = \cup_i Y_i^\wedge$.

The only not obvious fact is the inclusion $(\cup Y_i)^\wedge \subset \cup Y_i^\wedge$. But, if $x \in (\cup Y_i)^\wedge$, then for some $y \in \cup Y_i$, say $y \in Y_j$, y lies in the closure of $\{x\}$. But then $x \in Y_j^\wedge$, proving the assertion.

Note that the inclusion in (3) may be strict, even for a finite index set I. Indeed, consider the following example. Let R be arbitrary and let m be a maximal ideal of R. Put $Y = \mathrm{Spec}(R) - \{m\}$, then Y is generically closed since Y is open. If $Z = \{m\}$, then $Z^\wedge = \{p \in \mathrm{Spec}(R); p \subset m\}$. Now, we have

$$Y^\wedge \cap Z^\wedge = \{p \in \mathrm{Spec}(R); p \subset m, p \neq m\}$$

and if m is not also a minimal prime ideal of R, then it follows that $Y^\wedge \cap Z^\wedge \neq (Y \cap Z)^\wedge = \emptyset^\wedge = \emptyset$, showing that the inclusion may be strict! This example also shows that generically closed subsets are not necessarily open.

Of course, other examples may easily be given, e.g. let $X^{(n)}(R) \subset \mathrm{Spec}(R)$ consist of all primes p in R with $\mathrm{ht}(p) \leq n$, then $X^{(n)}(R)$ is clearly a generically closed subset of $\mathrm{Spec}(R)$.

(5.4.) Since the set of all generically closed subsets of a topological space X is closed under intersections also, the properties above show that these yield a topology on X which is finer than the original topology. This topology may have some weird properties, however. For instance, Spec(R) is in general not quasicompact for this topology, even if R is a Dedekind domain, say. Every closed subset V of X is the union of its irreducible components, i.e. maximal irreducible (closed) subsets of V. The irreducible closed sets of Spec(R) for this topology are all of the form $\{q \in \text{Spec}(R); q \not\subset p\}$ for some prime ideal p of R. One may give examples, where V is an infinite union of irreducible components, even when R is noetherian. Actually, it suffices to take R to be a noetherian domain with an infinite number of prime ideals of height 2 and to consider the collection $V = \{p \in \text{Spec}(R); \text{ht}(p) \geq 2\}$. Another bad property of this topology, is that quasicoherent sheaves do not behave decently. Indeed, if R is a Dedekind domain, then, associating pR_p to the "basic open set" $\{0, p\}$ if $p \neq 0$ and the field of fractions K of R to $\{0\}$ defines a quasicoherent sheaf on Spec(R), which is certainly not determined by its global sections.

(5.5.) Partitions of Spec(R) into subsets Y and V, where the first set is closed under generization (and hence the second under specialization, i.e. if $x \in V$ and y lies in the closure of $\{x\}$, then it follows that $y \in V$, or here : if $p \in V$ and $p \subset q$, then $q \in V$) also occur naturally, when considering abstract localization.

Let **L** be a filter of ideals of R, i.e. a collection of ideals with the following properties:

(a) if $L \in \mathbf{L}$ and L' is an ideal of R with $L \subset L'$, then $L' \in \mathbf{L}$;

(b) if $L, L' \in \mathbf{L}$, then $L \cap L' \in \mathbf{L}$, then we may associate to L a partition $(V_L = \mathrm{Spec}(R) \cap L, Y_L = \mathrm{Spec}(R) - V_L)$. Since L is a filter, clearly Y_L is closed under generization (and V_L under specialization).

One easily sees that every partition (V, Y) of $\mathrm{Spec}(R)$ into a subset V which is closed under specialization and a subset Y which is closed under generization, is of the form (V_L, Y_L) for some filter L. Indeed, just let L consist of all ideals L of R which contain a product $p_1. \ldots .p_n$, where $p_i \in V$ for $1 \leq i \leq n$, so $V = L \cap \mathrm{Spec}(R)$ with this definition, since of course $V \subset L$ and on the other hand, if $p_1. \ldots .p_n \subset p$ for some prime ideal p of R and some $p_i \in V$, then $p_i \subset p$ for some index i, hence $p \in V$, since V is closed under specialization.

(5.6.) In general localization theory, one is usually interested in some special type of filters, the so-called Gabriel filters or Gabriel topologies. Recall that L is said to be a <u>Gabriel filter</u> on R if it satisfies the following conditions:

(a) if $L \in L$ and $r \in R$, then $(L : r) \in L$;

(b) if L is an ideal of R and if for some $K \in L$ we have $(L : r) \in L$, for all $r \in K$, then I also belongs to L.

Here $(L : r)$ consists of all $s \in R$ with $sr \in L$!

Clearly not all filters are Gabriel filters, the filter constructed in the last part of the foregoing paragraph is actually a counterexample. However, Gabriel filters occur abundantly. Let us give some examples.

(a) <u>The Gabriel filter associated to an ideal</u>. Let I be a finitely generated ideal of R and consider the set $L(I)$ consisting of all ideals L of R containing a power I^n for some positive integer n. Clearly $L(I)$ is a filter. Moreover, let $L \in L(I)$, i.e. $I^n \subset L$ for some n, and let $r \in R$, then $I^n r \subset L$, hence $I^n \subset (L : r)$ and $(L : r) \in L(I)$, proving (a). Next, let L be an ideal of R and $K \in L(I)$, say $I^n \subset K$, with $(L : r) \in L(I)$ for all r in K, then, since I^n is finitely generated, we may find some positive integer p such that $I^{pr} \subset L$

for all $r \in I^n$, i.e. $I^{p+n} \subset L$ and $L \in \mathbf{L}(I)$. This proves that $\mathbf{L}(I)$ is a Gabriel filter.

(b) <u>The Gabriel filter associated to a prime ideal</u>. Let p be a prime ideal of R, then we may associate to it the set $\mathbf{L}(R - p)$ consisting of all ideals L of R with $L \not\subset p$. Of course $\mathbf{L}(R - p)$ is a filter, since e.g. $L \not\subset p$ and $L' \not\subset p$ yields $L \cap L' \not\subset p$ (as p is prime). Moreover, it is a Gabriel filter, since (a) if $L \not\subset p$ and $r \in R$, then $(L : r) \not\subset p$, since $L \subset (L : r)$ and (b) if L is an ideal of R with $(L : r) \not\subset p$ for all $r \in R$ with $I \not\subset p$, then $L \not\subset p$, as otherwise, with $r \in I - p$, we would have for all $s \in (L : r)$ that $sr \in L \subset p$, hence $s \in p$, i.e. $(L : r) \subset p$, against our assumption.

(c) <u>The Gabriel filter associated to the primes of height n</u>. Let $\mathbf{L}_n$ consist of all ideals L of R with $L \not\subset p$ for all $p \in \mathrm{Spec}^{(n)}(R)$, i.e. with $\mathrm{ht}(p) \leq n$. We leave it to the reader to check that this is again a Gabriel filter.

(5.7.) In fact, all of the foregoing are examples of the following construction. Let Y be a subset of $\mathrm{Spec}(R)$, then we associate to it the set $\mathbf{L}(Y)$ consisting of all ideals I of R such that $Y \subset X(I)$. Since $I' \subset I$ implies $X(I') \subset X(I)$ and since $X(I \cap J) = X(I) \cap X(J)$, clearly $\mathbf{L}(Y)$ is a filter and the fact that it is a Gabriel filter is checked exactly as for $\mathbf{L}(R - p)$. We then have:

(5.7.1.) $\mathbf{L}(I) = \mathbf{L}(X(I))$;

(5.7.2.) $\mathbf{L}(R - p) = \mathbf{L}(\{p\})$;

(5.7.3.) $\mathbf{L}_n = \mathbf{L}(\mathrm{Spec}^{(n)}(R))$.

Indeed, (2) and (3) are completely trivial, so, let us verify (1). If $L \in \mathbf{L}(X(I))$, then $X(I) \subset X(L)$ by definition, so $I \subset \mathrm{rad}(L)$, hence $I^n \subset L$ for some positive integer n, since I is finitely generated, proving that $\mathbf{L}(X(I)) \subset \mathbf{L}(I)$. Conversely, if $L \in \mathbf{L}(I)$, then $I^n \subset L$, hence $\mathrm{rad}(I) \subset \mathrm{rad}(L)$ and $X(I) \subset X(L)$, i.e. $L \in \mathbf{L}(X(I))$.

Since obviously $\mathbf{L}(Y) = \mathbf{L}(Y^\wedge)$ for any subset Y of Spec(R), we may restrict ourselves to subsets Y of Spec(R), which are generically closed. We thus have associated to any generically closed subset Y of Spec(R) a Gabriel filter $\mathbf{L}(Y)$ and conversely, to any Gabriel filter $\mathbf{L}$ on R a generically closed subset $Y_{\mathbf{L}} = \text{Spec}(R) - \mathbf{L}$ of Spec(R). Of course, one may wonder whether these data determine each other. Answers are given by the following results.

(5.8.) Lemma If Y is a generically closed subset of Spec(R), then $Y_{\mathbf{L}(Y)} = Y$.

Proof If $p \in Y_{\mathbf{L}(Y)}$, then $p \notin \mathbf{L}(Y)$, hence $Y \not\subset X(p)$. It follows that we may find some $q \in Y$ with $q \notin X(p)$, i.e. with $p \subset q$. Hence $p \in Y$, since Y is assumed to be generically closed. Since all of the above implications are equivalences, this also proves the converse.

Note that the foregoing proof actually shows that $Y_{\mathbf{L}(Y)} = Y^\wedge$ for any subset Y of Spec(R).

In general we do not not have $\mathbf{L} = \mathbf{L}(Y_{\mathbf{L}})$, unfortunately:

(5.9.) Lemma Let $\mathbf{L}$ be a Gabriel topology on R, then the following assertions are equivalent:

(5.9.1.) $\mathbf{L} = \mathbf{L}(Y)$ for some $Y \subset \text{Spec}(R)$;

(5.9.2.) $\mathbf{L} = \mathbf{L}(Y_{\mathbf{L}})$;

(5.9.3.) for every ideal $I \notin \mathbf{L}$, there is some prime ideal $I \subset p$ with $p \notin \mathbf{L}$.

Proof Obvious from the definitions and the foregoing.

(5.10.) Lemma Let $\mathbf{L}$ be a Gabriel topology on R,

(5.10.1.) if I is maximal with respect to the property $I \notin \mathbf{L}$, then I is a prime ideal, hence $I \in Y_{\mathbf{L}}$;

(5.10.2.) if $\mathbf{L}$ has a (filter) basis consisting of finitely generated ideals, then for every $I \notin \mathbf{L}$, we may find $p \in Y_{\mathbf{L}}$ such that $I \subset p$.

Proof (1) If $s, t \notin I$, then $I + Rs$ and $I + Rt$ must both belong to **L**, hence so does $(I + Rs)(I + Rt)$, since for all $x \in I + Rs$ we obtain that $I + Rt \subset ((I + Rs)(I + Rt) : x)$. But $(I + Rs)(I + Rt) \subset I + Rst$, hence $st \notin I$!

(2) Since $I \notin$ **L**, Zorn's Lemma yields an ideal $J \supset I$ which is maximal with respect to $J \notin$ **L**, since **L** has a basis of finitely generated ideals. It suffices to take $p = J$.

Gabriel topologies or filters possessing a basis of finitely generated ideals will occur frequently in the sequel; we will say that they are of <u>finite type</u>. If Y is a generically closed subset of $Spec(R)$, then we will say that $Spec(R)$ is of <u>finite type with respect to</u> Y if **L**(Y) is of finite type. More generally, if X is a scheme and Y a generically closed subset of X, then we call X <u>locally of finite type with respect to</u> Y, if we may find an open affine covering $\{Spec(R_a) = U_a; a \in A\}$ of X, such that **L**$(U_a \cap Y)$ is of finite type on R_a for all $a \in A$. We say that X is of <u>finite type with respect to</u> Y, if it is locally of finite type with respect to Y and quasicompact.

(5.11.) Corollary Every Gabriel topology of finite type on R is of the form **L** $=$ **L**(Y) for some $Y \subset Spec(R)$, where we may take $Y = Y_{\mathbf{L}}$.

(5.12.) Gabriel filters of the form **L**(Y) for some $Y \subset Spec(R)$ behave particularly well, because they allow local global arguments, in the sense that torsion may be checked locally at primes in Y.

Recall that to any Gabriel filter **L** we may associate a so-called Serre subcategory, cf. [Ga, Mi], denoted by $T_{\mathbf{L}}$ and consisting of all **L**-torsion R-modules. Here an R-module is said to be <u>L-torsion</u> if for all $m \in M$ we may find an ideal L in **L** such that $Lm = 0$. If **L** $=$ **L**(Y), then equivalently $M \in T_{\mathbf{L}}$ or $M_p = 0$ for all $p \in Y$. Indeed, if $M \in T_{\mathbf{L}}$ and $\mu = m/s \in M_p$ for some $p \in Y$, then $Im = 0$ for some $I \in$ **L**(Y), i.e. such that $Y \subset X(I)$. In particular, $I \not\subset p$, so $R_p m = R_p Im/s = 0$, hence $\mu = 0$.

Conversely, let M be an R-module with the property that $M_p = 0$ for all $p \in$ Y. If $m \in$ M, then we may find for each $p \in$ Y an $s_p \notin$ p such that $s_p m = 0$ (as $M_p = 0$!), hence with $I = \sum Rs_p$, we find $Im = 0$. But, $I \in$ **L**(Y), since $I \not\subset$ p for all $p \in$ Y, hence $Y \subset X(I)$. This proves the assertion.

It thus follows that an R-module M is torsion at **L**(Y) if and only if it is torsion at all $p \in$ Y, i.e.

$$T_L(Y) = \cap_{p \in Y} T_{L(R - p)}.$$

(5.13.) Let **L** be a Gabriel topology, then we will usually denote the quotient category R-mod/T_L by (R, **L**)-mod. It is well-known to be a Giraud subcategory of R-mod, cf. [St, VV1].

This category may also be realized in a more intuitive way from the point of view of idempotent kernel functors. Let us recall some of the machinery involved, and refer to the literature for more details.

For any R-module M, denote by $\sigma_L M$ or σM the submodule of M consisting of all m in M such that $Im = 0$ for some $I \in$ **L**. Associating σM to M defines an <u>idempotent kernel functor</u> in R-mod i.e. a left exact subfunctor of the identity in R-mod with the property that $\sigma(M/\sigma M) = 0$ for every $M \in$ R-mod. We call M a σ-<u>torsion</u> (resp. σ-<u>torsionfree</u>) R-module whenever $\sigma M = M$ (resp. $\sigma M = 0$). The idempotent kernel functor σ_L completely determines **L** and conversely. Indeed, to any idempotent kernel functor σ we may associate the Gabriel filter **L**(σ) consisting of the ideals I of R with the property that R/I is σ-torsion. It then follows that **L**(σ_L) = **L** and $\sigma_{L(\sigma)} = \sigma$, i.e. there is a one-to-one correspondence between Gabriel filters and idempotent kernel functors. In particular, this allows us to speak of idempotent kernel functors of <u>finite type</u>, i.e. such that the associated Gabriel filter **L**(σ) is of finite type.

Let E be an R-module, then we say that E is σ-<u>injective</u> (or σ-<u>divisible</u>) if for each exact sequence

$$0 \to M' \to M \to M'' \to 0$$

in R-mod, with $\sigma M'' = M''$ and every R-linear $f : M' \to E$, there exists a $g : M \to E$ extending f. We say that E is σ-<u>closed</u> (or <u>faithfully</u> σ-<u>injective</u>) if the g thus obtained is unique as such. Since σ is completely determined by its Gabriel filter $\mathbf{L}(\sigma)$, one should be able to check this property on $\mathbf{L}(\sigma)$ alone and, indeed, one easily shows that E is σ-closed if and only if every $f : I \to E$ with I in $\mathbf{L}(\sigma)$ uniquely extends to some $g : R \to E$. One may also prove that E is σ-closed if and only if E is σ-injective and σ-torsionfree.

The class of all σ-closed R-modules forms a full subcategory of R-mod, which we denote by (R, σ)-mod and call the <u>quotient category</u> of R-mod with respect to σ. One may prove that (R, σ)-mod and $(R, \mathbf{L}(\sigma))$-mod may be identified.

(5.14.) Let $i_\sigma : (R, \sigma)$-mod $\to$ R-mod denote the canonical inclusion. A left adjoint $\mathbf{a}_\sigma : $ R-mod $\to (R, \sigma)$-mod may be constructed as follows. First, let us call a σ-injective R-module E containing some R-module M a σ-<u>injective hull</u> of M if E/M is σ-torsion. Every σ-torsionfree R-module M possesses an essentially unique injective hull, which may be obtained from its injective hull E(M) : in the exact sequence

$$0 \to M \to E(M) \xrightarrow{\pi} E(M)/M \to 0,$$

we let $E = p^{-1}(\sigma(E(M)/M))$ and it is now easily checked that E is a σ-injective hull of M. More generally, let

$$E_\sigma(M) = \lim \mathrm{Hom}_R(L, M),$$

where M is an arbitrary R-module and where L runs through the ideals of R in $\mathbf{L}(\sigma)$, then one may show that if M is σ-torsionfree, then $E_\sigma(M)$ is "the" injective hull of M.

For any $M \in$ R-mod, put $\mathbf{a}_\sigma(M) = E_\sigma(M/\sigma M)$, then this yields a functor $\mathbf{a}_\sigma$: R-mod $\to (R, \sigma)$-mod, (since $E_\sigma(M/\sigma M)$ is σ-closed for any M and we may thus apply the universal properties!), which is easily seen to be an exact left adjoint to the inclusion i_σ : (R, σ)-mod $\to$ R-mod. In other words, (R, σ)-mod is a Giraud subcategory of R-mod.

More generally, it may be shown that <u>any</u> (strict) Giraud subcategory of R-mod may be realized this way. This is essentially the Gabriel - Popescu Theorem [Ga, GP, St].

(5.15.) The functor $Q_\sigma = i_\sigma \mathbf{a}_\sigma$: R-mod $\to$ R-mod is very important from the point of view of localization theory. In fact, for any R-module M we call $Q_\sigma(M)$ the <u>module of quotients</u> of M at σ. If S is a multiplicatively closed subset of R and if we define σ_S by letting $\sigma_S M$ consist of all $m \in$ M such that $sm = 0$ for some s in S, for any R-module M, then it is easy to see that σ_S actually defines an idempotent kernel functor in R-mod. The associated Gabriel topology $\mathbf{L}(S)$ consists of all ideals I of R such that $S \cap I \neq \emptyset$. In this case, it is easy to prove that with $\sigma = \sigma_S$ we have $Q_\sigma(M) = S^{-1}M$, the usual module of quotients of M at the multiplicatively closed subset S of R.

If $S = R - p$, then we obtain an idempotent kernel functor $\sigma_{R - p}$ and clearly $\mathbf{L}(\sigma_{R - p}) = \mathbf{L}(R - p)$. For any prime ideal p we thus have (with $\sigma = \sigma_{R - p}$) that $Q_\sigma(M) = M_p$, the module of quotients at p. More generally, we will usually write $Q_Y(M)$ for the module of quotients at the idempotent kernel functor σ_Y associated to the Gabriel filter $\mathbf{L}(Y)$.

The functor Q_σ may also be described directly by $Q_\sigma(M) = E_\sigma(E_\sigma(M))$ or, alternatively, by $Q_\sigma(M) = \lim \text{Hom}_R(I, M/\sigma M)$, where I runs through the ideals in $\mathbf{L}(\sigma)$. It follows easily that $Q_\sigma(R)$ possesses a canonical ring structure, induced by that of R and that $Q_\sigma(M) \in Q_\sigma(R)$-mod for any R-module M. From the universal properties enjoyed by Q_σ, it easily follows that if $f : M \to N$ is a morphism in R-mod such that Ker(f) and Coker(f) are σ-torsion, then f induces an isomorphism $Q_\sigma(f) : Q_\sigma(M) \to Q_\sigma(N)$. If σ is of finite type, then σ is completely determined by the set $Y_{L(\sigma)} = \{p \in \mathrm{Spec}(R); p \notin L(\sigma)\}$. One denotes this set by $K(\sigma)$ and one may prove that it consists of all prime ideals p of R such that R/p is σ-torsionfree. It follows from the foregoing that for any such σ, a morphism $f : M \to N$ induces an isomorphism $Q_\sigma(f)$ if and only if for each $p \in K(\sigma)$, it induces an isomorphism $f_p : M_p \to N_p$.

The idempotent kernel functor determined by $L(I)$, for some finitely generated ideal I of R, will be denoted by σ_I, the associated localization functor by Q_I. With these notations we then obtain:

(5.16.) Proposition Let I be an ideal of R and M an R-module. If I is finitely generated, then $\Gamma(X(I), \underline{Q}_M) = Q_I(M)$.

Proof If R is noetherian, then this is just Deligne's formula, cf. (3.9.). The proof we give here is easier, but assumes some techniques from localization theory.

Since I is finitely generated, we may find $s_1, \ldots, s_n \in R$ generating I, hence such that $X(s_1) \cup \ldots \cup X(s_n) = X(I)$. As in (3.9.), we write M_i for the localization of M at s_i, etc. hence, since $\underline{Q}_M$ is a sheaf, we obtain an exact sequence

$$(*) \qquad 0 \to \Gamma(X(I), \underline{Q}_M) \to \oplus_i M_i \to \oplus_{i,j} M_{ij},$$

where the maps are the obvious ones. Let $1 \le k \le n$, then by localizing at $s(k) = s_k$ we get an exact sequence

$$(**) \qquad 0 \to \Gamma(X(I), \underline{Q}_M)_{s(k)} \to \oplus_i M_{ik} \to \oplus_{i,j} M_{ijk},$$

On the other hand, using the covering $\{X(s_i) \cap X(s_k); 1 \le i \le n\}$ of $X(s_k)$, we get an exact sequence

$$(**) \qquad 0 \to M_k \to \oplus_i M_{ik} \to \oplus_{i,j} M_{ijk},$$

where the map $\oplus_i M_{ik} \to \oplus_{i,j} M_{ijk}$ coincides with the one in $(**)$. It follows that the canonical map $u : M \to \Gamma(X(I), \underline{Q}_M)$ yields for each s_k an isomorphism of the form $u_k : M_k \to \Gamma(X(I), \underline{Q}_M)_{s(k)}$, hence for each $p \in X(I)$, by localizing an isomorphism $u_p : M_p \to \Gamma(X(I), \underline{Q}_M)_p$. To finish the proof, it thus suffices to verify that $\Gamma(X(I), \underline{Q}_M)$ is σ_I-closed, but this follows immediately from the exactness of $(*)$ and the fact that each M_i is σ_I-closed (as $\sigma_I \le \sigma_i$, where σ_i is associated to the multiplicative system generated by s_i).

(5.17.) Although we have $Q_\sigma(M) = E_\sigma(E_\sigma(M))$ for any R-module M and $E_\sigma(M) = \lim \mathrm{Hom}_R(I, M)$, the explicit description of Q_σ as $Q_\sigma(M) = \lim \mathrm{Hom}_R(I, M/_\sigma M)$ involves dividing out the torsion of M at σ. One may wonder of course, when $E_\sigma(M) = Q_\sigma(M)$ for all M (and not only for M torsionfree!). The answer is given by looking at so-called stable Gabriel filters or idempotent kernel functors.

Recall that if M is an R-module, then any Gabriel filter **L** defines a linear topology on M, the so-called **L**-topology, by choosing as a fundamental system of neighborhoods of $0 \in M$ the set of all R-modules $M' \subset M$, such that M/M' is **L**-torsion. If M' is a submodule of M, then the **L**-topology on M induces on M' a subspace (= induced) topology, which is in general weaker than the **L**-topology on M'. The following result is due to Gabriel [Ga].

(5.18.) Proposition If **L** is a Gabriel topology on R, then the following assertions are equivalent:

(5.18.1.) for every module M and every submodule M' of M, the **L**-topology on M' coincides with the subspace topology induced from the **L**-topology on M;

(5.18.2.) the class of **L**-torsion modules is closed under injective hulls;

(5.18.3.) every injective R-module is split by its **L**-torsion submodule.

(5.19.) We call **L** or its associated idempotent kernel functor σ_L <u>stable</u> if it satisfies one of the equivalent properties of (5.18.). We claim that if σ is stable, then for any R-module M we have $Q_\sigma(M) = E_\sigma(M)$.

Indeed, consider the exact sequence $0 \to \sigma M \to M \to M/\sigma M \to 0$, which induces the exact sequence

$$0 \to \lim \mathrm{Hom}_R(I, \sigma M) \to \lim \mathrm{Hom}_R(I, M) \to \lim \mathrm{Hom}_R(I, M/\sigma M) \to$$
$$\to \lim \mathrm{Ext}_R^1(I, \sigma M).$$

Let us show that both extreme terms vanish. First, if $f : I \to \sigma M$ represents an element of $\lim \mathrm{Hom}_R(I, \sigma M)$, then we claim that $\mathrm{Ker}(f) \in \mathbf{L}(\sigma)$, which will prove that the class of f in the inductive limit vanishes, since it coincides with that of $0 = f|\mathrm{Ker}(f) : \mathrm{Ker}(f) \to \sigma M$. So, pick $i \in I$, then we may find $L_i \in \mathbf{L}(\sigma)$ with $L_i f(i) = 0$. If $J = \sum L_i i$, where i runs through I, then certainly $J \subset \mathrm{Ker}(f)$. Moreover, for all $i \in I$ we have $L_i \subset (J : i)$, hence $(J : i) \in \mathbf{L}(\sigma)$, so $J \in \mathbf{L}(\sigma)$ and hence $\mathrm{Ker}(f) \in \mathbf{L}(\sigma)$, indeed.

On the other hand, if E is an injective hull of σM, then E is a σ-torsion module, since σ is assumed to be stable. Moreover, the exact sequence $0 \to \sigma M \to E \to E/\sigma M \to 0$ induces an exact sequence

$$\lim \mathrm{Hom}_R(I, E/\sigma M) \to \lim \mathrm{Ext}_R^1(I, \sigma M) \to 0.$$

By the same argument as before, $\lim \mathrm{Hom}_R(I, E/\sigma M) = 0$, so we find that $\lim \mathrm{Ext}_R^1(I, \sigma M) = 0$, indeed.

This proves that $E_\sigma(M) = \lim \mathrm{Ext}_R^1(I, M) = \lim \mathrm{Ext}_R^1(I, M/\sigma M) = Q_\sigma(M)$!

Note that since Krull's Theorem clearly implies condition (5.18.1.) for any idempotent kernel functor σ_I associated to an ideal I of a noetherian ring R, clearly (3.8.) directly follows from the foregoing remarks, since σ_I is then stable. This also explains why (5.17.) generalizes (3.9.).

(5.20.) Lemma Let σ be an idempotent kernel functor of finite type in R-mod and p a prime ideal of R, then for any R-module M there is a canonical isomorphism $Q_\sigma(M)_p = Q_\sigma(M_p)$.

Proof If I is a finitely generated ideal of R, then $X(I) = \cup_i X(g_i)$ for some finite number of elements $g(i) = g_i$ of R, so taking sections of the quasicoherent sheaf associated to M_p on Spec(R), we get an exact sequence

$$0 \to Q_I(M_p) \to \oplus_i (M_p)_{g(i)} \to \oplus_{i,j}(M_p)_{g(i)g(j)}.$$

We get a similar sequence for $\underline{Q}_M$, and these fit into a commutative exact diagram

$$0 \to Q_I(M)_p \to \oplus_i (M_{g(i)})_p \to \oplus_{i,j}(M_{g(i)g(j)})_p.$$
$$\downarrow \qquad\qquad \downarrow$$
$$0 \to Q_I(M_p) \to \oplus_i (M_p)_{g(i)} \to \oplus_{i,j}(M_p)_{g(i)g(j)}.$$

(with obvious maps), hence $Q_I(M)_p = Q_I(M_p)$.

Since obviously $M_p/\sigma M_p = M_p/(\sigma M)_p = (M/\sigma M)_p$ (see also (5.21.)), to prove that $Q_\sigma(M)_p = Q_\sigma(M_p)$, we may clearly assume M to be σ-torsionfree. For any finitely generated $I \in L(\sigma)$ we then have an injection $Q_I(M) \to Q_\sigma(M)$ (induced by $\sigma_I \leq \sigma$!) and it follows that $Q_\sigma(M) = \cup Q_I(M)$, where I runs through the finitely generated ideals in $L(\sigma)$. So, $Q_\sigma(M)_p = \cup Q_I(M)_p$ and similarly $Q_\sigma(M_p) = \cup Q_I(M_p)$. Since $Q_I(M)_p = Q_I(M_p)$ for all I, we find $Q_\sigma(M)_p = Q_\sigma(M_p)$, indeed.

(5.21.) Lemma Assume λ and μ are idempotent kernel functors of finite type in R-mod, then $Q_\lambda Q_\mu = Q_\mu Q_\lambda$.

Proof First we claim that for any R-module N we have $Q_\lambda(\mu N) = \mu Q_\lambda(N)$.

Indeed, let $n \in Q_\lambda(\mu N)$, then for some finitely generated $I \in \mathbf{L}(\lambda)$, we have that $In \subset \mu N/\lambda(\mu N)$. As for each $i \in I$, we may find $n_i \in \mu N$ such that $in = \tilde{n}_i$ mod $\lambda(\mu N)$, we may pick for any such i an ideal $J_i \in \mathbf{L}(\mu)$ such that $J_i in = J_i \tilde{n}_i = (J_i n_i)^\sim = 0$, hence a single $J \in \mathbf{L}(\mu)$ such that $JIn = 0$, as I is finitely generated. It follows that $Jn \subset \lambda Q_\lambda(\mu N) = 0$, hence $n \in \mu Q_\lambda(N)$.

Conversely, if $n \in \mu Q_\lambda(N)$, then we may find finitely generated ideals $I \in \mathbf{L}(\lambda)$ and $J \in \mathbf{L}(\mu)$ such that $In \subset N/\lambda N$ and $Jn = 0$. For $i \in I$, choose $n_i \in N$ such that $in = \tilde{n}_i$ mod λN; as $Jn = 0$, we find $(Jn_i)^\sim = 0$, i.e. $Jn_i \subset \lambda N$ and as J is finitely generated, there is an $I'_i \in \mathbf{L}(\lambda)$ with $JI'_i n_i = I'_i Jn_i = 0$, hence $I'_i n_i \subset \mu N$. It follows that $I'_i \tilde{n}_i \subset \mu N/\lambda(\mu N)$, hence $I'_i in \subset \mu(N/\lambda(\mu N))$ and $in \in Q_\lambda(\mu N)$. We obtain $In \subset Q_\lambda(\mu N)$ and so, finally, that $n \in Q_\lambda(\mu N)$. This proves the assertion.

Consider the following diagram

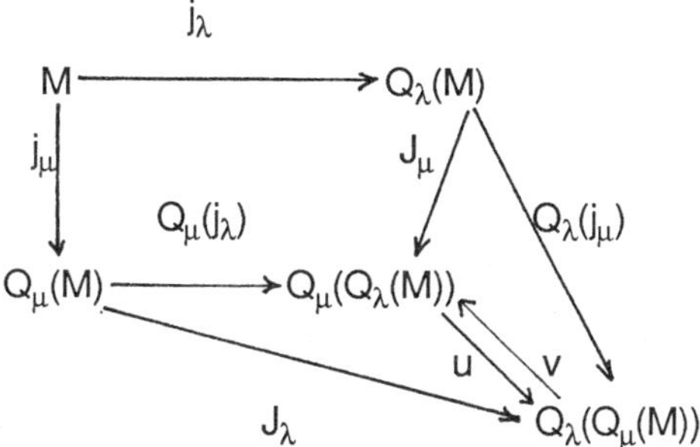

First, consider $f = Q_\mu(j_\lambda) : Q_\mu(M) \to Q_\mu(Q_\lambda(M))$. Localizing f at any $p \in \mathbf{K}(\lambda)$ and applying (5.20.) yields that f_p is an isomorphism, so Ker(f) and Coker(f) are λ-torsion, hence there exists a unique morphism $u : Q_\mu(Q_\lambda(M)) \to Q_\lambda(Q_\mu(M))$ extending J_λ. In a similar way, there exists a unique $v : Q_\lambda(Q_\mu(M)) \to Q_\mu(Q_\lambda(M))$ extending J_μ (working with $g = Q_\lambda(j_\mu)$).

An easy uniqueness argument then shows that u and v are mutually inverse to each other, whence the assertion.

(5.22.) Lemma If σ and τ are idempotent kernel functors of finite type in R-mod, then $\mathbf{L}(\mathbf{K}(\sigma) \cap \mathbf{K}(\tau))$ consists of all ideals L of R, containing a product IJ, where $I \in \mathbf{L}(\sigma)$ and $J \in \mathbf{L}(\tau)$.

Proof Using the fact that σ and τ have finite type, it easily follows that

$$\mathbf{L} = \{L < R;\ \exists\, I \in \mathbf{L}(\sigma),\ \exists\, J \in \mathbf{L}(\tau),\ IJ \subset L\}$$

is a Gabriel filter of finite type. Now, for any prime ideal p of R, we have $p \in \mathbf{L}$ if and only if for all $K \in \mathbf{K}(\sigma) \cap \mathbf{K}(\tau)$ we have $I \not\subset p$, i.e. $p \in \mathbf{K}(\sigma) \cap \mathbf{K}(\tau)$. It follows that $\mathbf{L} = \mathbf{L}(Y_L) = \mathbf{L}(\mathbf{K}(\sigma) \cap \mathbf{K}(\tau))$.

(5.23.) Corollary Let σ and τ be as in the previous Lemma and denote by $Q_{\sigma\tau}$ the localization associated to $\mathbf{L}(\mathbf{K}(\sigma) \cap \mathbf{K}(\tau))$, then $Q_{\sigma\tau} = Q_\sigma Q_\tau$.

Proof Let M be an R-module, then we claim that $Q_\sigma Q_\tau(M) = Q_\tau Q_\sigma(M)$ is $\mathbf{L}(\mathbf{K}(\sigma) \cap \mathbf{K}(\tau))$-injective. Indeed, first pick $I \in \mathbf{L}(\sigma)$ and $J \in \mathbf{L}(\tau)$, then any morphism $f : IJ \to Q_\sigma Q_\tau(M)$ extends to $f' : J \to Q_\sigma Q_\tau(M)$, since J/IJ is clearly σ-torsion and since $Q_\sigma Q_\tau(M)$ is σ-closed. Now f' extends to a morphism $g : R \to Q_\sigma Q_\tau(M)$ since $J \in \mathbf{L}(\tau)$ and $Q_\sigma Q_\tau(M)$ is also τ-closed. To prove that any $f : L \to Q_\sigma Q_\tau(M)$ with $L \in \mathbf{L}(\mathbf{K}(\sigma) \cap \mathbf{K}(\tau))$ extends to R, it suffices to restrict f first to some IJ with $I \in \mathbf{L}(\sigma)$ and $J \in \mathbf{L}(\tau)$. We then obtain a map $g : R \to Q_\sigma Q_\tau(M)$ as previously. But g|L = f, since if h = g|L - f, then h|L = 0, hence h factorizes through L/IJ by a map $h' : IJ/L \to Q_\sigma Q_\tau(M)$. But $L/IJ \subset R/IJ$, so L/IJ is $\mathbf{L}(\mathbf{K}(\sigma) \cap \mathbf{K}(\tau))$-torsion, i.e. h' = 0, which proves the assertion.

It is also clear that $Q_\sigma Q_\tau(M)$ is torsionfree with respect to $\mathbf{L}(\mathbf{K}(\sigma) \cap \mathbf{K}(\tau))$, since for all $I \in \mathbf{L}(\sigma)$, $J \in \mathbf{L}(\tau)$ and $m \in Q_\sigma Q_\tau(M)$ we have that IJm = 0 implies Jm = 0, since $Q_\sigma Q_\tau$ is σ-torsionfree, hence m = 0 since it is also τ-

torsionfree. Finally, consider the map $j : M \to Q_\sigma Q_\tau(M)$, which is the composition of $u : M \to Q_\tau(M)$ and $v : Q_\tau(M) \to Q_\sigma Q_\tau(M)$. For each $p \in \mathbf{K}(\sigma)$, the map v_p is an isomorphism, so the morphism $j_p : M_p \to Q_\sigma Q_\tau(M)_p$ is just the map $u_p : M_p \to Q_\tau(M)_p$. It follows that if we have $p \in \mathbf{K}(\tau)$, then j_p is an isomorphism, since u_p is, i.e. $\mathrm{Ker}(j)_p = \mathrm{Coker}(j)_p = 0$ for all $p \in \mathbf{K}(\sigma) \cap \mathbf{K}(\tau)$, hence $\mathrm{Ker}(j)$ and $\mathrm{Coker}(j)$ are $\mathbf{L}(\mathbf{K}(\sigma) \cap \mathbf{K}(\tau))$-torsion. It follows that $Q_{\sigma\tau}(M) = Q_{\sigma\tau}(Q_\sigma Q_\tau(M)) = Q_\sigma Q_\tau(M)$.

We have seen in the previous Section that any continuous map $f : X \to Y$ between topological spaces induces a functor $f_p^{-1} : \underline{P}(Y) \to \underline{P}(X)$ and a functor $f^{-1} : \underline{S}(Y) \to \underline{S}(X)$. The functor f_p^{-1} is determined by associating to a presheaf of abelian groups Q on Y the presheaf of abelian groups $f_p^{-1}Q$ on X, given by putting $f_p^{-1}Q(U) = \lim Q(V)$, where U is open in X and where V runs through the open subsets of Y with $U \subset V$. The functor $f^{-1} : \underline{S}(Y) \to \underline{S}(X)$ is then defined as $\mathbf{a}f_p^{-1}i : \underline{S}(Y) \to \underline{S}(X)$, where $i : \underline{S}(Y) \to \underline{P}(Y)$ is the inclusion and $\mathbf{a} : \underline{P}(X) \to \underline{S}(X)$ is the sheafification functor.

As usually, if X is a subset of Y with the induced topology and f is the inclusion of X into Y, then we write $Q|X$ for $f^{-1}Q$. We also write $\Gamma(Y, E)$ for $\Gamma(Y, E|Y)$.

(5.24.) Proposition Let Y be a generically closed subset of $\mathrm{Spec}(R)$ and let M be an R-module. If $\mathrm{Spec}(R)$ is of Y-finite type, then there is a canonical isomorphism

$$\Gamma(Y, \underline{Q}_M) = Q_Y(M),$$

where Q_Y denotes the localization functor with respect to $\sigma_{\mathbf{L}(Y)}$.

Proof Let $f : Y \to \mathrm{Spec}(R)$ denote the canonical inclusion and write $Y(I) = X(I) \cap Y = f^{-1}(X(I))$ for any ideal I of R. Since the topology on Y is induced from $\mathrm{Spec}(R)$, every open subset of Y is of this form, hence the $Y(I)$ with I

finitely generated form a basis B for the topology on Y. Now, $f_p^{-1}\underline{Q}_M$ on Y is given by $(f_p^{-1}\underline{Q}_M)(Y(I)) = \lim \underline{Q}_M(X(J))$, where $X(J) \supset Y(I)$, i.e. $J \in \mathbf{L}(Y(I))$ and if we choose I finitely generated, then by (5.18.) we may restrict ourselves to J's that are finitely generated too, i.e. $f_p^{-1}\underline{Q}_M(Y(I)) = \lim Q_J(M)$, where J runs through the finitely generated ideals of $\mathbf{L}(Y(I))$. For each such J there is an obvious map $Q_J(M) \to Q_{Y(I)}(M)$, hence a morphism $t_I : \lim Q_J(M) \to Q_{Y(I)}(M)$. We may define a presheaf of $\underline{Q}_M|Y$-modules $\underline{H}_M$ on B, by putting $\underline{H}_M(Y(I)) = Q_{Y(I)}(M)$ for each finitely generated ideal I of R. Pick $p \in Y$, then the stalk of $\underline{H}_M$ at p is given by

$$\underline{H}_{M,p} = \lim Q_{Y(I)}(M) = \lim Q_I Q_Y(M) = Q_Y(M)_p = M_p,$$

where $Y(I)$ and $X(I)$ contain p, and where we have used (5.21.) and (5.23.). It thus follows that the t_I yield a presheaf morphism $t : f_p^{-1}\underline{Q}_M \to \underline{H}_M$ on B, which is isomorphic in all $p \in Y$.

It now suffices to show that $\underline{H}_M$ is a sheaf (on B) to finish the proof. Indeed, t then induces an isomorphism $f^{-1}\underline{Q}_M = \underline{H}_M$ and we then clearly obtain that $\Gamma(Y, \underline{Q}_M) = \Gamma(Y, f^{-1}\underline{Q}_M) = \Gamma(Y, \underline{H}_M) = Q_Y(M)$. Choose an open covering $\{Y(I_\alpha); \alpha \in A\}$ of some $Y(I)$, all of these open sets lying in B, and assume $m \in \underline{H}_M(Y(I))$ is mapped to 0 by all $\mathrm{res}_\alpha : \underline{H}_M(Y(I)) \to \underline{H}_M(Y(I_\alpha))$, then by definition $m \in \sigma_{Y(I)}(Q_{Y(I)}(M)) = 0$.

On the other hand, assume that for all $\alpha \in A$, there is given some $m_\alpha \in \underline{H}_M(Y(I_\alpha)) = Q_{Y(\alpha)}(M)$, where $Y(\alpha) = Y(I_\alpha)$, such that $m_\alpha|Y(I_\alpha) \cap Y(I_\beta) = m_\beta|Y(I_\alpha) \cap Y(I_\beta)$. Since $Y(I) = \cup\, Y(I_\alpha)$, obviously $Y(I) \subset X(\Sigma\, I_\alpha)$, i.e. $\Sigma\, I_\alpha \in \mathbf{L}(Y(I))$ by definition. But, since $Y(I)$ is of finite type by (5.22.), it follows that for some finitely generated $L \in \mathbf{L}(Y(I))$ we have $L \subset \Sigma\, I_\alpha$, i.e. we may find a finite number of ideals $I_1, ..., I_n$ amongst the I_α such that $K = \Sigma_{i \leq n} I_i \in \mathbf{L}(Y(I))$. Hence $Y(I) \subset X(K) = X(I_1) \cup ... \cup X(I_n)$, so $Y(I)$ is covered by $\{Y(I_1), ..., Y(I_n)\}$. Since $\underline{Q}_M$ is a sheaf on Spec(R), there is an exact sequence

$$0 \to \Gamma(X(K), \underline{Q}_M) \to \oplus_i \Gamma(X(I_i), \underline{Q}_M) \to \oplus_{i,j} \Gamma(X(I_i I_j), \underline{Q}_M)$$

with obvious morphisms, which reduces to

$$0 \to Q_K(M) \to \oplus_i Q_{I(i)}(M) \to \oplus_{i,j} Q_{I(i)I(j)}(M),$$

where $I(i) = I_i$, ... as usually.

Localizing at σ_Y yields an exact sequence

$$0 \to Q_Y Q_K(M) \to \oplus_i Q_Y Q_{I(i)}(M) \to \oplus_{i,j} Q_Y Q_{I(i)I(j)}(M),$$

Now, by (5.22.), $Q_Y Q_K(M) = Q_{Y(K)}(M) = Q_{Y(I)}(M)$ and similarly $Q_Y Q_{I(i)}(M) = Q_{Y(i)}(M)$, ..., so the above sequence reduces to

$$0 \to \Gamma(Y(K), \underline{H}_M) \to \oplus_i \Gamma(Y(I_i), \underline{H}_M) \to \oplus_{i,j} \Gamma(Y(I_i I_j), \underline{H}_M)$$

We thus find some $m \in \underline{H}_M(Y(I))$ with $m|Y(I_i) = m_i$ for $1 \le i \le n$.

Finally, for any $\alpha \in A$, let $n_\alpha = m|Y(I_\alpha)$, then for each $1 \le i \le n$ we have $n_\alpha|Y(I_\alpha) \cap Y(I_i) = (m|Y(I_\alpha))|Y(I_\alpha) \cap Y(I_i) = (m|Y(I_i))|Y(I_\alpha) \cap Y(I_i) = m_i|Y(I_i) \cap Y(I_\alpha) = m_\alpha|Y(I_i) \cap Y(I_\alpha)$, hence $m|Y(I_\alpha) = m_\alpha$, since $Y(I_\alpha)$ is covered by the $Y(I_\alpha) \cap Y(I_i)$.

This shows that $\underline{H}_M$ is a sheaf and finishes the proof.

(5.25.) If σ is stable or M is σ-torsionfree, then $f_p^{-1}\underline{Q}_M = f^{-1}\underline{Q}_M$, i.e. we have $Q_\sigma(M) = \lim Q_I(M)$. Indeed, first note that $t : \lim Q_I(M) \to Q_\sigma(M)$ is injective in both cases, for if $\mu \in \lim Q_I(M)$ and $t(\mu) = 0$, then there exists a finitely generated $I \in L(\sigma)$ and some $m \in Q_I(M)$ such that $m \in \mathrm{Ker}(Q_I(M) \to Q_\sigma(M))$ and such that m represents μ. Now, by the description of the map $Q_I(M) \to Q_\sigma(M)$ it then follows that $m \in \sigma Q_I(M)$, hence that $Jm = 0$ for some finitely generated $J \in L(\sigma)$. Let m' be the image of m in the module $Q_{IJ}(M)$, then m' still represents μ. Moreover, we have $Jm' = 0$ in $Q_{|J}(M)$, hence $m' \in \sigma_{|J}Q_{|J}(M) = 0$, i.e. $\mu = 0$ and t is injective. Of course, this is trivial if M is torsionfree at σ.

On the other hand, if σ is stable, then every $\mu \in Q_\sigma(M)$ is represented by some R-linear $m : I \to M$, where I is finitely generated in $\mathbf{L}(\sigma)$. Let $n : I \to M \to M/\sigma_|M$ represent $\tilde{n} \in \lim Q_I(M)$, then $t(\tilde{n})$ represents the same element as m, proving that t is surjective in this case. The same argument (but with $\sigma_|M = 0$!) works if M is σ-torsionfree.

(5.26.) Fortunately (for us), situations where σ is of finite type occur rather frequently. For example, if R is noetherian, then <u>every</u> idempotent kernel functor in R-mod is of finite type. However, the noetherian assumption is usually too restrictive for our purposes. Instead, let us introduce the notion of σ-noetherian rings. Let σ be an idempotent kernel functor in R-mod and M an R-module, then we say that M is σ-<u>noetherian</u> if $Q_\sigma(M)$ is a noetherian object in (R, σ)-mod, i.e. if it satisfies the ascending chain condition. We call the ring itself σ-noetherian if it is σ-noetherian as an R-module.

As we have pointed out in [VV3] e.g., if R is σ-noetherian, then the idempotent kernel functor σ is of finite type. Indeed, if $L \in \mathbf{L}(\sigma)$, pick $l_1 \in L$ and let $L_1 = Rl_1$. If there exists $l_2 \in L - Rl_1$, let $L_2 = Rl_1 + Rl_2$, etc. We then obtain that $L = Rl_1 + \dots + Rl_n$, for some n or else we have an ascending chain $Q_\sigma(L_1) \subset \dots \subset Q_\sigma(L) = Q_\sigma(R)$. Since R is σ-noetherian, $Q_\sigma(L_m) = Q_\sigma(R)$ for some m, hence $L_m \in \mathbf{L}(\sigma)$ and $L_m \subset L$.

As in [VV2], call an R-module M σ-<u>finitely generated</u> resp. σ-<u>finitely presented</u> if we may find a finitely generated R-submodule N of M such that M/N is σ-torsion resp. if we may find a finitely presented R-module N together with an R-linear $u : N \to M$ with σ-torsion kernel and cokernel.

For details concerning these notions, we refer to [VV2, VV3]. Let us only recall here the following results, which we include for completeness' sake :

(5.27.) Lemma Let σ be an idempotent kernel functor in R-mod, then

(5.27.1.) an R-module M is σ-noetherian if and only if every $N \subset M$ is σ-finitely generated;

(5.27.2.) if R is σ-noetherian, then every σ-finitely generated R-module is σ-finitely presented.

Proof (1) If M and N are as in the statement, choosing an increasing chain of finitely generated R-submodules N' of N yields some such N' with $Q_\sigma(N') = Q_\sigma(N)$ since M is σ-noetherian by assumption. Hence N/N' is σ-torsion, so N is σ-finitely generated. For the converse, note that an ascending chain $\{M_\alpha; \alpha \in A\}$ of submodules of $Q_\sigma(M)$ in (R, σ)-mod may be realized by a chain $\{N_\alpha; \alpha \in A\}$ of submodules of M with $Q_\sigma(N_\alpha) = M_\alpha$. Choose a finitely generated $N_1 \subset N = \cup N_\alpha$ such that $Q_\sigma(N_1) = Q_\sigma(N)$, then for some $\alpha' \in A$ we have $N_1 \subset N_{\alpha'}$ and one easily deduces that $M_\alpha = Q_\sigma(N_\alpha) \subset Q_\sigma(N_{\alpha'})$ and one easily deduces that $M_\alpha = Q_\sigma(N_\alpha) \subset Q_\sigma(N_{\alpha'}) = M_{\alpha'}$ for all α, i.e. the chain $\{M_\alpha; \alpha \in A\}$ stops at α'.

(2) If M is finitely generated, i.e. there exists a surjective map $p : R^n \to M$ for some positive integer n, then Ker(p) is σ-finitely generated, since R^n is σ-noetherian if R is, hence M is easily seen to be σ-finitely presented. Next, if M is σ-finitely generated, then by the first part, there exists a finitely presented R-module P and a morphism $u : P \to N$ with σ-torsion kernel and cokernel, where $N \subset M$ has the property that M/N is σ-torsion. The composition $P \to N \subset M$ does the job!

Our main interest in σ-finitely presented modules stems from the following result:

(5.28.) Lemma Let σ be an idempotent kernel functor in R-mod and let M, M' be R-modules. If M is σ-finitely presented, M' is σ-closed and either S is a multiplicative subset of R with $\mathbf{L}(\sigma) \subset \mathbf{L}(S)$ or S is arbitrary, but σ is of finite type, there is a canonical isomorphism

$$S^{-1}\mathrm{Hom}_R(M, M') \cong \mathrm{Hom}_R(S^{-1}M, S^{-1}M').$$

Proof Choose an R-linear map $u : N \to M$ with N finitely presented and with σ-torsion kernel and cokernel, then the fact that M' is σ-closed implies that the induced morphism $\mathrm{Hom}_R(M, M') \to \mathrm{Hom}_R(N, M')$ is an isomorphism. Since N is finitely presented, it follows that $S^{-1}\mathrm{Hom}_R(N, M') = \mathrm{Hom}_R(S^{-1}N, S^{-1}M')$, hence it suffices to show that the second member reduces to $\mathrm{Hom}_R(S^{-1}M, S^{-1}M')$.

If $\mathbf{L}(\sigma) \subset \mathbf{L}(S)$, then we easily see that $S^{-1}Q_\sigma = Q_\sigma S^{-1} = S^{-1}$ and if σ is of finite type, (5.16.) yields that $S^{-1}Q_\sigma = Q_\sigma S^{-1}$ as well, since $\mathbf{L}(S)$ is obviously also of finite type. In any case, we thus find that $S^{-1}M'$ is σ-closed. Moreover, we also have that $Q_\sigma(S^{-1}N) = S^{-1}Q_\sigma(N) \cong S^{-1}Q_\sigma(M) = Q_\sigma(S^{-1}M)$, the isomorphism being induced by u. So, $\mathrm{Hom}_R(S^{-1}N, S^{-1}M') \cong \mathrm{Hom}_R(Q_\sigma(S^{-1}N), S^{-1}M') \cong \mathrm{Hom}_R(Q_\sigma(S^{-1}M), S^{-1}M') \cong \mathrm{Hom}_R(S^{-1}M, S^{-1}M')$.
(The first and the last isomorphism are induced by the universal property of Q_σ!).

(5.29.) Let Y be a generically closed subset of a scheme X and assume that X is covered by affine open subsets $\{U_\alpha; \alpha \in A\}$, say $U_\alpha = \mathrm{Spec}(R_\alpha)$ for some ring R_α, then for every $\alpha \in A$, the set $Y_\alpha = Y \cap U_\alpha$ is generically closed and induces an idempotent kernel functor sa in R_α-mod. We call X <u>locally Y-noetherian</u> if X may be covered by open affines as above, such that for each index α the ring R_α is σ_α-noetherian. If X is quasicompact and locally Y-noetherian, then we say that X is <u>Y-noetherian</u>. Equivalently, if it may be covered by finitely many open affines $\mathrm{Spec}(R_\alpha)$- such that each R_α is σ_α-noetherian. If Y is an arbitrary subset of X, then we say that X is (locally) Y-noetherian if it is (locally) $Y^\wedge$-noetherian.

(5.30.) Proposition Let Y be a subset of X, then X is locally Y-noetherian if and only if for every open subset U = Spec(R) of X, the ring R is $\sigma_{Y \cap U}$ noetherian.

Proof By definition, one implication is obvious. Let us write σ for $\sigma_{Y \cap U}$, then we have to prove that R is σ-noetherian. One easily reduces the problem to the following: if σ is an idempotent kernel functor in R-mod and if $f(1) = f_1, ..., f(r) = f_r$ are elements of R which generate the unit ideal and such that $R_{f(\alpha)}$ is σ-noetherian for each $1 \leq \alpha \leq r$, show that R is σ-noetherian. Now, first note that with these notations, for every ideal I of R we have $I = \cap_\alpha j_\alpha^{-1}(j_\alpha(I)R_{f(\alpha)})$, where $j_\alpha : R \to R_{f(\alpha)}$ is the localizing morphism. Let I be an ideal of R and let $\{I_t; t \in T\}$ be an ascending chain of finitely generated ideals with $\cup \{I_t; t \in T\} = I$, then for each $1 \leq \alpha \leq r$, we have that $\{j_\alpha(I_t)R_{f(\alpha)}; t \in T\}$ is an ascending chain of finitely generated $R_{f(a)}$-ideals with $\cup j_\alpha(I_t)R_{f(\alpha)} = j_\alpha(I)R_{f(\alpha)}$, hence, since $R_{f(\alpha)}$ is σ-noetherian and the α are only finite in number, there exists an index $t \in T$ such that $R_{f(\alpha)}/j_\alpha(I_t)R_{f(\alpha)}$ is σ-torsion for all $1 \leq \alpha \leq r$. Let $J = I_t$, then we obtain $R/J = R/\cap_\alpha j_\alpha^{-1}(j_\alpha(J)R_{f(\alpha)}) \subset \oplus R/j_\alpha^{-1}(j_\alpha(J)R_{f(\alpha)})$, which is σ-torsion since we know that each $R/j_\alpha^{-1}(j_\alpha(J)R_{f(\alpha)})$ injects into $R_{f(\alpha)}/j_\alpha(J)R_{f(\alpha)}$. This finishes the proof.

(5.31.) Lemma Let σ be an idempotent kernel functor in R-mod, such that R is σ-torsion, then $\mathbf{K}(\sigma) \subset \mathrm{Spec}(R)$ is quasicompact.

Proof This has essentially been proved in (5.24.)

(5.32.) Corollary Let Y be a generically closed subset of X for which X is Y-noetherian, then Y is quasicompact.

(5.33.) Lemma Let I be an ideal of R such that $X(I) \subset \mathrm{Spec}(R)$ is noetherian, then $\mathrm{Spec}(R)$ is X(I)-noetherian.

Proof Since $X(I)$ is noetherian, it may be covered by a finite number of affines $\{X(f(i)); 1 \leq i \leq n\}$ with $f(i) \in R$, such that $R_{f(i)}$ is noetherian for each i. Moreover, since $X(I) = X(\Sigma Rf(i))$, we may assume that I is finitely generated (by these $f(i)$!). Let K be an arbitrary ideal of R, then we have to show that K is σ_I-finitely generated. Now, if $j_i : R \to R_{f(i)}$ denotes the localizing morphism, then by assumption $j_i(K)R_{f(i)}$ is a finitely generated $R_{f(i)}$-ideal, hence we may find a finitely generated R-ideal $L \subset K$ such that $j_i(L)R_{f(i)} = j_i(K)R_{f(i)}$ for all i. It remains to check that K/L is σ_I-torsion to finish the proof. This may be seen as follows : pick $k \in K$, then we may find a positive integer p such that $j_i(k) = j_i(l_i)/f(i)^p$ for all i and some $l_i \in L$. In particular, since this yields that $j_i(f(i)^p k - l_i) = 0$, we may find some q such that for all i we have $f(i)^q(f(i)^p k - l_i) = 0$, so $f(i)^{p+q} k \in L$ for all i, hence $I^r k \subset L$ for some sufficiently large positive integer r. This implies that K/L is σ_I-torsionfree and finishes the proof.

(5.34.) It now easily follows that if U is an open subset of the scheme X, then X is U-noetherian if and only if X is quasicompact and U is a noetherian scheme. Indeed, from (5.32.) we know that if Y is a generically closed subset of X, then X being Y-noetherian implies that Y is quasicompact. This already implies that if X is U-noetherian, then U is certainly quasicompact. Now, let $X(f)$ be an open affine subset of $U = X(I)$, where R is assumed to be σ_I-noetherian, then since $X(f) \subset X(I)$, we have that R is $X(f)$-noetherian, i.e. R_f is noetherian or equivalently, the scheme $X(f)$ is noetherian. It follows that U is noetherian as well, as it may be covered by a finite number of open affine subschemes of this form. From this, the global version follows easily.

Conversely, let X be quasicompact and U a noetherian open subscheme, then we want to show that X is U-noetherian. One easily reduces to the affine case, i.e. $X = Spec(R)$ and $U = X(I)$ for some ideal I of R. So, (5.33.) finishes the proof in this case!

6 EXTENDING SHEAVES

(6.1.) In this Section we will study the behaviour of quasicoherent and coherent sheaves on generically closed subsets of a scheme X with respect to extensions to the whole scheme. In order to make things run smoothly, we will have to put some (harmless) finiteness assumptions on X. For example, we will assume X to be locally of finite type with respect to Y or locally Y-noetherian, depending on whether we work with quasicoherent or coherent sheaves.

The results obtained in this direction depend heavily on the behaviour of (quasi)coherent sheaves on affine schemes. It thus appears that some generalities on perfect localizations in the sense of [Ga] or [Gold] are needed in order to derive the statements below. This explains why we start off with a subsection on property (T) and its applications. Note also that along the way we will strengthen some related results in [GD].

(6.2.) Let us start by recollecting some well known results on perfect localizations. Recall that if S is a multiplicatively closed subset of R, then for any R-module M we have $S^{-1}M = S^{-1}R \otimes_R M$. On the other hand, if σ is an arbitrary idempotent kernel functor in R-mod, then we do not necessarily have $Q_\sigma(M) = Q_\sigma(R) \otimes_R M$. If this condition holds for any M, then as in [Gold] we say that σ has property (T) or as in [Ga] that it induces a perfect localization.

The following result presents some conditions equivalent to property (T) :

(6.3.) Proposition [Gold] For any idempotent kernel functor σ in R-mod, the following assertions are equivalent :

(6.3.1.) $Q_\sigma(M) = Q_\sigma(R) \otimes_R M$ for any R-module M;

(6.3.2.) $j_\sigma(I)Q_\sigma(R) = Q_\sigma(R)$ for any $I \in \mathbf{L}(\sigma)$, where $j_\sigma : R \to Q_\sigma(R)$ is the canonical morphism;

(6.3.3.) every $Q_\sigma(R)$-module is σ-closed as an R-module;

(6.3.4.) every $Q_\sigma(R)$-module is σ-torsionfree as an R-module;

(6.3.5.) the functor Q_σ is exact and commutes with direct sums.

(6.4.) It has been pointed out in [Gold] that Q_σ is right exact (hence exact) if and only if R is σ-projective, i.e. if it has the following property : let $I \in L(\sigma)$ and assume $p : M \to M''$ is a surjective morphism between σ-torsionfree R-modules, then for any R-linear map $f : I \to M$, we may find some $J \in L(\sigma)$ with $J \subset I$ and some R-linear $g : J \to M$, such that $f|J = pg$, i.e. such that the following square commutes :

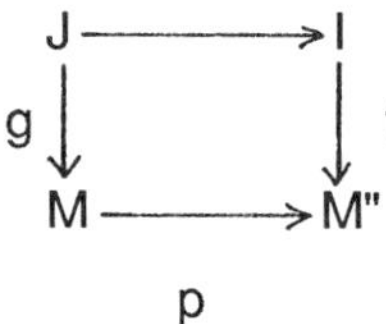

Note also that Q_σ is exact if and only if the canonical inclusion of (R, σ)-mod into R-mod is exact. Indeed, if $0 \to M' \to M \to^\pi M'' \to 0$ is exact in (R,σ)-mod, then this means that $0 \to M' \to M \to M''$ is an exact sequence of σ-closed R-modules in R-mod and that T, the cokernel of the map π, is σ-torsion. But, applying the exact functor Q_σ to $M \to M'' \to T \to 0$ yields that $M \to M'' \to 0$ is exact, since T is σ-torsion and M, M'' are σ-closed, hence π is surjective as well!

(6.5.) The other condition in (6.4.5.), i.e. that Q_σ commutes with direct sums also has an elegant equivalent description. Indeed, it has been proved in [Gold] that this condition is equivalent to σ being a noetherian idempotent kernel functor, i.e. if $\{I_\alpha; \alpha \in A\}$ is a countable chain of ideals of R with $\cup I_\alpha \in L(\sigma)$, then $I_\alpha \in L(\sigma)$ for some $\alpha \in A$.

It is clear that if σ has property (T), then σ has finite type. Indeed, if $I \in$ $\mathbf{L}(\sigma)$, then $j_\sigma(I)Q_\sigma(R) = Q_\sigma(R)$, so we may write $1 = \Sigma j_\sigma(i_\alpha)r_\alpha$ for some finite number of $i_\alpha \in I$ and $r_\alpha \in R$. Let $J = \Sigma Ri_\alpha \subset I$, then it suffices to check that $J \in \mathbf{L}(\sigma)$ to finish the proof of the assertion. Pick $K \in \mathbf{L}(\sigma)$ such that $Kr_\alpha \subset$ $j_\sigma(R)$ for all α, then we have $j_\sigma(K) = Kj_\sigma(1) = K\Sigma j_\sigma(i_\alpha)r_\alpha \subset \Sigma j_\sigma(i_\alpha)j_\sigma(R) = j_\sigma(J)$, so we find $K \subset j_\sigma^{-1}(j_\sigma(K)) \subset J$ and thus $J \in \mathbf{L}(\sigma)$, indeed.

If σ has finite type, then obviously σ is noetherian, for if $\{I_\alpha; \alpha \in A\}$ is a countable ascending chain of ideals in R with $\cup_\alpha I_\alpha \in \mathbf{L}(\sigma)$, then we may find a finitely generated ideal $J \in \mathbf{L}(\sigma)$ with $J \subset \cup_\alpha I_\alpha$. But then $J \subset I_\alpha$ for some $\alpha \in A$ (each of the <u>finite</u> number of generators of J lies in some I_α!) and $I_\alpha \in \mathbf{L}(\sigma)$.

Note that one actually easily proves that if Q_σ is exact, then σ is of finite type if and only if σ is noetherian.

(6.6.) Note In view of the foregoing remark, one may wonder whether any noetherian idempotent kernel functor σ also has finite type. The following counterexample, taken from [Ru] is due to G. Bergman. We include it here for completeness' sake.

Let G be a nondiscrete ordered group, in which the identity e is not the greatest lower bound of any countable set of elements greater that e (e.g. let G be an uncountable product of the integers, ordered lexicographically). Let K be a field and $S = \{g \in G; g > e\}$. In $R = K[S]$, the semigroup algebra of S over K, let I be the ideal of R generated by all $g \in$ S, then I is maximal and $I^2 = I$, i.e. there is some idempotent kernel functor σ in R-mod with $\mathbf{L}(\sigma) = \{I, R\}$. Let $J_1 \subset \ldots \subset J_n \subset \ldots$ be a countable chain of ideals with $J_i \neq I$, R for all indices i, then for each i there exists some $g \neq e$ in G with $g \leq g_i$ for all i. But then $g \notin \cup_i J_i$ and $\cup_i J_i \in \mathbf{L}(\sigma)$. It follows that σ is noetherian, but I does not contain any finitely generated $J \in \mathbf{L}(\sigma)$.

(6.7.) We will show below that property (T) is a local property. We will need some preparatory remarks first.

If $f : R \to S$ is a ringmorphism, then for any idempotent kernel functor σ in R-mod, we may define an idempotent kernel functor $f_*\sigma$ in S-mod by putting $(f_*\sigma)(M) = \sigma(_RM)$ for any S-module M, where $_RM$ is M viewed as an R-module through f.

Clearly for any ideal L of S we have $L \in \mathbf{L}(f_*\sigma)$ if and only if $f^{-1}(L) \in \mathbf{L}(\sigma)$. Indeed, if $L \in \mathbf{L}(f_*\sigma)$, then S/L is σ-torsion, hence so is $R/f^{-1}(L) \subset S/L$, so it follows that $f^{-1}(L)$ belongs to $\mathbf{L}(\sigma)$.

Conversely, if $f^{-1}(L) \in \mathbf{L}(\sigma)$ for some ideal L of S, then denoting the R-module structure on S induced by f with . , we have $f^{-1}(L).1 = f(f^{-1}(L)) \subset L$, i.e. S/L is annihilated by $f^{-1}(L)$, hence S/L is σ-torsion and hence $f_*\sigma$-torsion.

If Spec(R) is partitioned by σ into $(\mathbf{K}(\sigma), \mathbf{L}(\sigma) \cap \mathrm{Spec}(R))$, then the corresponding partition of Spec(S) is $(\mathbf{K}(f_*\sigma), \mathbf{L}(f_*\sigma) \cap \mathrm{Spec}(S))$, where for any prime ideal q of S we have $q \in \mathbf{K}(f_*\sigma)$ if and only if $f^{-1}(q) \in \mathbf{K}(\sigma)$ and hence similarly for the complement.

(6.8.) If $p \in \mathrm{Spec}(R)$ and $j_p : R \to R_p$ denotes the localizing morphism, then for any σ we denote $j_{p,*}\sigma$ by $\sigma(p)$. We then have that $\mathbf{L}(\sigma(p))$ consists of all ideals of R_p of the form $L_p = LR_p$ for some $L \in \mathbf{L}(\sigma)$. Indeed, if $L \in \mathbf{L}(\sigma)$, then $j_p\text{-}1(L_p) \supset L$, hence $j_p\text{-}1(L_p) \in \mathbf{L}(\sigma)$ and $L_p \in \mathbf{L}(\sigma(p))$. Conversely, if $K \in \mathbf{L}(\sigma(p))$, then $L = j_p\text{-}1(K)$ belongs to $\mathbf{L}(\sigma)$ and $K = L_p$. We also have the following result:

(6.9.) Lemma Let p be a prime ideal of R and σ an idempotent kernel functor of finite type in R-mod, then for any R-module M we have

$$Q_\sigma(M)_p = Q_{\sigma(p)}(M_p).$$

Proof Since σ and $\sigma_{R\text{-}p}$ both have finite type, using (5.16.) one easily deduces that the canonical morphism $j_\sigma : M \to Q_\sigma(M)$ yields a morphism $j_{\sigma,p} : M_p \to Q_\sigma(M)_p$, with $\sigma(p)$-torsion kernel and cokernel. To finish the proof, it thus suffices to check that $Q_\sigma(M)_p$ is $\sigma(p)$-closed. Now, trivially $\sigma(p)(Q_\sigma(M)_p) = \sigma(Q_\sigma(M)_p) = \sigma Q_\sigma(M_p) = 0$, so we only have to verify that $Q_\sigma(M)_p$ is $\sigma(p)$-injective. Let $f : L_p \to Q_\sigma(M_p) = Q_\sigma(M)_p$ be an arbitrary R_p-linear map, where $L \in \mathbf{L}(\sigma)$, hence where L_p is arbitrary in $\mathbf{L}(\sigma(p))$ and write $j_{L,p} : L \to L_p$ resp. $j_{R,p} : R \to R_p$ for the localization morphisms, then there exists a (unique) R-linear $g : R \to Q_\sigma(M_p)$ such that $g|L = fj_{L,p}$, since $Q_\sigma(M_p)$ is σ-closed. On the other hand, since $Q_\sigma(M_p)$ is also $\sigma_{R\text{-}p}$-closed, the morphism $g : R \to Q_\sigma(M_p)$ extends to some $h : R_p \to Q_\sigma(M_p)$. It thus follows that $(h|L_p)j_{L,p} = hj_{R,p}|L = g|L = fj_{L,p}$, hence by a unicity argument, that $h|L_p = f$. This proves the assertion.

(6.10.) Proposition If σ is an idempotent kernel functor of finite type in R-mod, then the following assertions are equivalent:

(6.10.1.) σ has property (T) in R-mod;

(6.10.2.) $\sigma(p)$ has property (T) in R_p-mod, for every $p \in \mathrm{Spec}(R)$.

Proof If σ has property (T) and M is an R_p-module, then we have

$$Q_{\sigma(p)}(R_p) \otimes M = Q_\sigma(R)_p \otimes M = Q_\sigma(M)_p = Q_{\sigma(p)}(M_p),$$

hence $\sigma(p)$ has property (T). This proves (1) $\Rightarrow$ (2).

To prove the other implication, assume $\sigma(p)$ has property (T) in R_p-mod, for all $p \in \mathrm{Spec}(R)$, let $M \in$ R-mod and look at the multiplication map

$$u : Q_\sigma(R) \otimes M \to Q_\sigma(M),$$

then for any $p \in \mathrm{Spec}(R)$ we obtain an induced map

$$u_p : Q_{\sigma(p)}(R_p) \otimes M_p = (Q_\sigma(R) \otimes M)_p \to Q_\sigma(M)_p = M_p,$$

which is an isomorphism for every p, since $\sigma(p)$ is assumed to have property (T). Hence u is an isomorphism, proving that σ has property (T).

(6.11.) Note The second statement (6.10.2.) in the previous Proposition is equivalent to $\sigma(p)$ having property (T) for all $p \in \text{Spec}(R) \cap \mathbf{L}(\sigma)$. Indeed, since for any $p \in \mathbf{K}(s)$ we have $Q_{\sigma(p)}(M_p) = Q_\sigma(M)_p = M_p$, it is clear that for all of these p, the idempotent kernel functor $\sigma(p)$ trivially has property (T).

(6.12.) Recall from (4.9.) that if R is a commutative ring and $(X, \underline{O}_X)$ an arbitrary scheme, then there is a functorial isomorphism

$$[X, \text{Spec}(R)] \cong [R, \Gamma(X, \underline{O}_X)]$$

given by taking global sections. The inverse image is given as follows. Let $u : R \to \Gamma(X, \underline{O}_X)$ be a ring morphism and $x \in X$, then there is a diagram

$$R \to \Gamma(X, \underline{O}_X) \to \underline{O}_{X,x} \supset m_x,$$

where $j_x : \Gamma(X, \underline{O}_X) \to \underline{O}_{X,x}$ is the stalk map and m_x the unique maximal ideal of the local ring $\underline{O}_{X,x}$. The map $^a u : X \to \text{Spec}(R)$ on the base spaces, corresponding to u, is given by sending $x \in X$ to $(j_x u)^{-1}(m_x)$ in $\text{Spec}(R)$ and the map on the sheaf level is easily deduced from this.

Now assume that I is a finitely generated ideal of R such that X(I) in $\text{Spec}(R)$ is affine, i.e. such that

$$(X(I), \underline{O}_R | X(I)) \cong (\text{Spec}(S), \underline{O}_S)$$

for some ring S, which is then easily seen to be isomorphic to $\Gamma(X(I), \underline{O}_R)$

$= Q_I(R)$. So, the map on global sections is an isomorphism. Assume that $q \in X(I)$ and denote by $j_q : \Gamma(X(I), \underline{Q_R}) \to R_q$ the stalk map (which is easily seen to reduce to the localization map $Q_I(R) \to Q_I(R)_q = R_q$, then it follows from the foregoing that the isomorphism on base spaces $X(I) \cong \mathrm{Spec}(Q_I(R))$ the isomorphism is given by sending $q \in X(I)$ to $(j_q)^{-1}(m_q) \in \mathrm{Spec}(Q_I(R))$, where m_q is the maximal ideal of R_q and this map is bijective.

Let n be a positive integer and assume that $j_I(I^n)Q_I(R) \neq Q_I(R)$, where the map $j_I : R \to Q_I(R)$ is the localization map. Let p be a prime ideal of $Q_I(R)$ containing $j_I(I^n)Q_I(R)$ and assume q is the prime ideal in $X(I)$ corresponding to p, i.e. $p = (j_q)^{-1}(m_q)$. We then find that

$$j_q(j_I(I^n)Q_I(R)) \subset j_q((j_q)^{-1}(m_q)) \subset m_q,$$

hence $I_q \subset m_q$. But, since $q \in X(I)$, we have $I_q = R_q$, which yields a contradiction. It thus follows that $j_I(I^n)Q_I(R) = Q_I(R)$, which easily yields that σ_I has property (T).

Let us call $Y \subset X$ an <u>affine subset</u> of the locally ringed space $(X, \underline{Q_X})$ if $(Y, \underline{Q_X}|Y)$ is affine. We then have the following result, which generalizes the foregoing remarks:

(6.13.) Proposition If σ is an idempotent kernel functor of finite type in R-mod, then the following assertions are equivalent:
(6.13.1.) K(σ) is affine in $\mathrm{Spec}(R)$;
(6.13.2.) σ has property (T);
(6.13.3.) K$(\sigma) = (j_\sigma)^{-1}(\mathrm{Spec}(Q_\sigma(R)))$, where $j_\sigma : R \to Q_\sigma(R)$ is the localization map.
Proof The proofs of $(1) \Rightarrow (3)$ and $(3) \Rightarrow (2)$ are straightforward generalizations of the proof given for $X(I) \subset \mathrm{Spec}(R)$, so it remains to

verify that (2) $\Rightarrow$ (1). More precisely, we will show below that if σ has property (T), then $(\mathbf{K}(\sigma), \underline{O}_R|\mathbf{K}(\sigma)) = (\text{Spec}(S), \underline{O}_S)$, where $S = Q_\sigma(R)$.

Indeed, if $p \in \mathbf{K}(\sigma)$, then we claim that $pQ_\sigma(R) \in \text{Spec}(Q_\sigma(R))$. Clearly $j_\sigma(p)$ is a prime ideal of $j_\sigma(R) = R/\sigma R$, for if $j_\sigma(x)j_\sigma(y) \in j_\sigma(p)$ for some $x, y \in R$, then $Kxy \subset p$ for some $K \in \mathbf{L}(\sigma)$ and it follows that $x \in p$ or $y \in p$, since obviously $K \not\subset p$.

So, we may assume R to be σ-torsionfree. Now, if $a, b \in Q_\sigma(R)$ and if ab lies in $pQ_\sigma(R)$, then we may find some $L \in \mathbf{L}(s)$ such that $La \subset P$, $Lb \subset R$ and $Lab \subset p$, hence for example $La \subset p$. But then property (T) implies that $a \in aQ_\sigma(R) = aLQ_\sigma(R) \subset pQ_\sigma(R)$, i.e. $pQ_\sigma(R)$ is prime, indeed. Moreover, property (T) also implies that $pQ_\sigma(R)$ is σ-closed and it easily follows that $pQ_\sigma(R) = Q_\sigma(p)$. From this one may derive that sending p to $Q_\sigma(p) = pQ_\sigma(R)$ yields a homeomorphism of $\mathbf{K}(\sigma)$ onto $\text{Spec}(Q_\sigma(R))$. Finally, since one easily checks that $I \in \mathbf{L}(Q_\sigma(R) - pQ_\sigma(R))$ if and only if $(j_\sigma)^{-1}(I) \not\subset p$, i.e. $(j_\sigma)^{-1}(I) \in \mathbf{L}(R - p)$, the idempotent kernel functor in $Q_\sigma(R)$-mod associated to $pQ_\sigma(R)$ is just $j_{\sigma,*}\sigma_{R\,-p}$. So, it follows that the localization of $Q_\sigma(R)$ at $pQ_\sigma(R)$ is isomorphic to R_p, whence one easily deduces the isomorphism of locally ringed spaces we need. This finishes the proof.

(6.14.) Corollary Let M be an R-module and σ an idempotent kernel functor of finite type in R-mod, then the following assertions are equivalent:

(6.14.1.) $\mathbf{K}(\sigma)$ is affine in $\text{Spec}(R)$;

(6.14.2.) $\Gamma(\mathbf{K}(\sigma), \underline{O}_M) = M \otimes \Gamma(\mathbf{K}(\sigma), \underline{O}_R)$ for all R-modules M;

(6.14.3.) the functor $M \to \Gamma(\mathbf{K}(\sigma), \underline{O}_M)$ is exact.

Proof If $\mathbf{K}(\sigma)$ is affine, then σ has property (T). Since $\Gamma(\mathbf{K}(\sigma), \underline{O}_M) = Q_\sigma(M)$ by (5.24.), it follows from (6.3.) that (1) $\Rightarrow$ (2). Next, since (2) says that σ has property (T), the implication (2) $\Rightarrow$ (3) is just the implication (6.3.1.) $\Rightarrow$ (6.3.5.). Finally, as (3) shows that Q_σ is exact and as our assumptions imply σ to be noetherian, the implication (3) $\Rightarrow$ (1) is exactly (6.13.) together with the description (6.3.5.) of property (T).

(6.15.) This result may be related to Serre's Theorem (cf. [Ha1] for example). Assume for simplicity's sake that R is noetherian, then any idempotent kernel functor σ in R-mod is stable and has finite type. We also claim that for any injective R-module E the morphism $j_\sigma : E \to Q_\sigma(E)$ is surjective. Indeed, from (5.20.) it follows that $Q_\sigma(E) = E_\sigma(E) = \lim [I, E]$, where I runs through $\mathbf{L}(\sigma)$. The map j_σ is given by sending any $e \in E$ to the map $f_e : I \to E : i \to ie$, where I is arbitrary in $\mathbf{L}(\sigma)$. Now, let $\alpha \in Q_\sigma(E)$ be represented by some $f : I \to E$, then this map extends to a $g : R \to E$, hence $f = f_e$, where $e = g(1)$, i.e. $\alpha = j_\sigma(e)$. This shows that j_σ is surjective. From this one may deduce that the sheaf $\underline{Q}_E|\mathbf{K}(\sigma)$ is flabby on $\mathbf{K}(\sigma)$ for any idempotent kernel functor σ in R-mod, since every open subset of $\mathbf{K}(\sigma)$ is of the form $\mathbf{K}(\tau)$ for some $\tau \geq \sigma$.

Let M be an arbitrary R-module, then M embeds in an injective R-module E. Let $T = E/M$, then we get an exact sequence of sheaves of $\underline{Q}_R|\mathbf{K}(\sigma)$-modules

$$0 \to \underline{Q}_M|\mathbf{K}(\sigma) \to \underline{Q}_E|\mathbf{K}(\sigma) \to \underline{Q}_T|\mathbf{K}(\sigma) \to 0,$$

hence, if the functor $M \to \Gamma(\mathbf{K}(\sigma), \underline{Q}_M)$ is exact, an exact sequence of R-modules

$$0 \to \Gamma(\mathbf{K}(\sigma), \underline{Q}_M) \to \Gamma(\mathbf{K}(\sigma), \underline{Q}_E) \to \Gamma(\mathbf{K}(\sigma), \underline{Q}_T) \to 0.$$

On the other hand, (2.21.) yields an exact sequence

$$0 \to \Gamma(\mathbf{K}(\sigma), \underline{Q}_M) \to \Gamma(\mathbf{K}(\sigma), \underline{Q}_E) \to \Gamma(\mathbf{K}(\sigma), \underline{Q}_T) \to H^1(\mathbf{K}(\sigma), \underline{Q}_M) \to 0$$

since the fact that $\underline{Q}_E|\mathbf{K}(\sigma)$ is flabby on $\mathbf{K}(\sigma)$ implies that $H^1(\mathbf{K}(\sigma), \underline{Q}_E) = 0$, by (2.30.). Comparing the last two sequences, we get $H^1(\mathbf{K}(\sigma), \underline{Q}_M) = 0$ for any $M \in$ R-mod. Conversely, if this holds, obviously $M \to \Gamma(\mathbf{K}(\sigma), \underline{Q}_M)$ is an exact functor. Hence we have proved:

(6.15.) Proposition If R is noetherian and if σ is an idempotent kernel functor in R-mod, then $\mathbf{K}(\sigma)$ is an affine subset of Spec(R) if and only if $H^1(\mathbf{K}(\sigma), \underline{Q}_M) = 0$ for all R-modules M.

Of course, one may prove as in [Ha1], that it is actually sufficient to verify the last condition for ideals of R only.

The following technical result will be needed below:

(6.16.) Lemma If Y is a generically closed subset of the scheme X, then X is locally of finite type with respect to Y if and only if for every open affine subset U = Spec(R) of X the idempotent kernel functor $\sigma_{U \cap Y}$ has finite type in R-mod.

Proof If Spec(A) is affine and $Y = \mathbf{K}(\sigma)$ for some idempotent kernel functor σ of finite type, then for any $f \in A$ we clearly have $X(f) \cong \mathrm{Spec}(A_f)$ and $\sigma_{X(f) \cap Y}$ has finite type in A_f-mod. It thus follows that if the scheme X is locally of finite type with respect to some generically closed subset Y of X, then X possesses a basis of open subsets $\{U(i); i \in I\}$, all of them affine and such that $\sigma_{U(i) \cap Y}$ has finite type in the category $\Gamma(U(i), \underline{Q}_X)$-mod.

Cover U = Spec(R) by a finite number of these, say $U = \cup_\alpha \{U(\alpha) = \mathrm{Spec}(R_\alpha)\}$, where $R_\alpha = R_{f(\alpha)}$ for some $f(\alpha) \in R$ and $\sigma_{Y \cap U(\alpha)}$ has finite type in R_α-mod. If I in $\mathbf{L}(Y \cap U)$, i.e. $Y \cap U \subset X(I)$, then $U(\alpha) \cap X(I) \supset Y \cap U(\alpha)$ for all indices α. Now $X(I) \cap U(\alpha)$ may be viewed as the set of all p in $\mathrm{Spec}(R_\alpha)$ such that $I_{f(\alpha)} \not\subset p$. But then we may find some $K_\alpha \subset I_{f(\alpha)}$ where $K_\alpha \in \mathbf{L}(Y \cap U(\alpha))$ and such that K_α is finitely generated in R_α-mod. Moreover, for each α we may then pick a finitely generated ideal J_α of R such that $J_\alpha \subset I$ and $J_\alpha R_\alpha = K_\alpha$. Let $J = \Sigma_\alpha J_\alpha \subset I$, then we claim that J belongs to $\mathbf{L}(Y \cap U)$. Indeed, otherwise $\cup_\alpha Y \cap U_\alpha = Y \cap U \not\subset X(J)$, i.e. for some index α we would have $Y \cap U_\alpha \not\subset X(J) \cap U_\alpha = X(JR_\alpha)$ and $X(J_\alpha R_\alpha)$ $\subset X(JR_\alpha)$. This contradiction finishes the proof of one implication. The other implication is obvious.

In the next result we will have to assume the scheme X to be separated. For the definition and properties of this notion, we refer the reader to the literature, cf. [GD, Ha1, ...]. We will only use the fact that if X is separated and U and V are open affines in X, then so is $U \cap V$. For the proof of this we refer to loc. cit.

(6.17.) Proposition Let X be a separated scheme and Y a generically closed subset of X such that X is locally of finite type with respect to Y. For any quasicoherent sheaf of $\underline{Q}_Y = \underline{Q}_X|Y$-modules E on Y, we may find a quasicoherent sheaf of $\underline{Q}_X$-modules F on X such that $F|Y = E$.

Proof By assumption, Y may be covered by open affines $U(i) = U_i = \mathrm{Spec}(R_i)$ of X such that, with $V(i) = V_i = Y \cap U_i$, there exists for each index i an exact sequence

$$(*) \qquad \underline{Q}_Y^{(I)}|V_i \to \underline{Q}_Y^{(J)}|V_i \to E|V_i \to 0,$$

for some idex sets I, J (depending on i!). It follows from (6.16.) that each idempotent kernel functor $\sigma(i) = \sigma_{Y \cap U(i)}$ in R_i-mod has finite type, hence also that $V_i = \mathbf{K}(\sigma(i))$.

From (5.24.) we deduce that $\Gamma(V_i, \underline{Q}_Y) = \Gamma(V_i, \underline{Q}_Y|V_i) = Q_{\sigma(i)}(R_i)$.

Taking sections over V_i, the exact sequence (*) thus yields a morphism

$$v_i : Q_{\sigma(i)}(R_i)^{(I)} \to Q_{\sigma(i)}(R_i)^{(J)},$$

as $Q_{\sigma(i)}$ commutes with direct sums, since $\sigma(i)$ has finite type, hence is noetherian, and we put $F(i) = Q_{\sigma(i)}(\mathrm{Coker}(v_i))$. Since for each $p \in V_i$ we have $\underline{O}_{Y,p} = (\underline{O}_X|Y)_p = \underline{O}_{X,p} = R_{i,p} = Q_{\sigma(i)}(R_i)_p$, it follows for the associated sheaf $\underline{O}_{F(i)}$ on U_i, that $\underline{O}_{F(i)}|V_i = E|V_i$.

We claim that the $\underline{O}_{F(i)}$ glue together well, i.e. for any pair of indices i and j, we have $\underline{O}_{F(i)}|U_i \cap U_j = \underline{O}_{F(j)}|U_i \cap U_j$. Indeed, let $U(ij) = U(i) \cap U(j)$, then U(ij) is affine, since X is separated. Hence for k = i or j, the subset U(ij) of U(k) corresponds to an idempotent kernel functor $\tau(k)$ in R_k-mod, such that $U(ij) = \mathbf{K}(\tau(k))$ and such that $\tau(k)$ induces a perfect localization.

Now, $\underline{O}_{F(k)}|V(k) \cap U(ij) = \underline{O}_E|V(k) \cap U(ij) = \underline{O}_E|Y \cap U(ij)$, so it follows that $\Gamma(V(i) \cap U(ij), \underline{O}_{F(i)}) = \Gamma(V(j) \cap U(ij), \underline{O}_{F(j)})$. But, using (5.22.), it follows that $\Gamma(V(k) \cap U(ij), \underline{O}_{F(k)}) = Q_{V(k) \cap U(ij)}(F(k)) = Q_{U(ij)}(Q_{V(k)}(F(k)) = Q_{\tau(k)}(Q_{\sigma(k)}(F(k))) = Q_{\tau(k)}(F(k)) = \Gamma(U(ij), \underline{O}_{F(k)})$, hence $\underline{O}_{F(i)}|U(ij) = \underline{O}_{F(j)}|U(ij)$ indeed, since U(ij) is affine. It thus follows that there exists some quasicoherent sheaf of $\underline{O}_U$-modules F' on $U = \cup_i U_i$ with $F|U_i = \underline{O}_{F(i)}$.

To conclude, pick an open affine covering $\{W(s) = W_s = \text{Spec}(R_s); s \in S\}$ of X, with $\sigma_{Y \cap W(s)}$ of finite type in R_s-mod for all $s \in S$, let $F'_s = F'|U \cap W_s$ for any $s \in S$ and denote by $\mu_s : U \cap W_s \to W_s$ resp $\mu_{is} : U_i \cap W_s \to W_s$ resp. $\mu_{ijs} : U(ij) \cap W_s \to W_s$ the canonical inclusions. Here, for each $s \in S$ we restrict to a finite number of indices i, corresponding to a <u>finite</u> number of subsets U_i covering W_s (using the fact that W_s is affine, hence quasicompact!)

Since X is separated, the subsets $U_i \cap W_s$ and $U(ij) \cap W_s$ are affine, hence the

$$\mu_{is,*}(F'|U_i \cap W_s) \quad \text{resp.} \quad \mu_{ijs,*}(F'|U(ij) \cap W_s)$$

are quasicoherent sheaves of $\underline{O}_{W(s)}$-modules on W_s, cf. (4.11.). Moreover, we have an exact sequence

$$0 \to \mu_{s,*}F'_s \to \oplus_i \mu_{is,*}(F'|U_i \cap W_s) \to \oplus_{i,j}\mu_{ijs,*}(F'|U(ij) \cap W_s)$$

and it follows that $\mu_{s,*}F'_s$ is a quasicoherent sheaf of $\underline{O}_{W(s)}$-modules on W_s as well.

We claim that $\mu_{s,*}F'_s|U \cap W_s = F'|U \cap W_s$. Indeed, for any $W \subset U \cap W_s$ yields

$$\lim \mu_{s,*}F'_s(W') = \lim F'_s(W' \cap U) = F'(W),$$

where W' is open in W_s, which yields the assertion. We thus obtain some sheaf $F''_s = \mu_{s,*}F'_s$ on W_s, which is quasicoherent and such that

$$F''_s|Y \cap W_s = F'|Y \cap W_s = E|Y \cap W_s.$$

As W_s is affine, F''_s is of the form $\underline{O}_G$, for some R_s-module G, and upto changing F''_s, we may obviously assume G to be $\sigma_{Y \cap W(s)}$-closed. Argueing as before, we derive that

$$F''_s|W_s \cap W_t = F''_t|W_s \cap W_t,$$

for any pair of indices $s, t \in S$, hence that the F''_s glue together well. We thus obtain a quasicoherent sheaf of $\underline{O}_X$-modules F on X such that $F|W_s = F''_s$, i.e. such that $F|Y \cap W_s = F''_s|Y \cap W_s = E|Y \cap W_s$, i.e. with the property that $F|Y = E$. This finishes the proof.

(6.18.) Corollary Let σ be an idempotent kernel functor of finite type in R-mod and let $Y = \mathbf{K}(\sigma)$, then there is a one-to-one correspondence between

(6.18.1.) σ-closed R-modules;

(6.18.2.) quasicoherent sheaves of $\underline{O}_Y$-modules on Y.

Proof First recall that if M and N are σ-closed R-modules such that $\underline{O}_M|Y = \underline{O}_N|Y$, then $M = N$ (it suffices to take global sections). Moreover, for any

R-module M we have $\underline{Q}_M|Y = \underline{Q}_N|Y$, since the map $j_\sigma : M \to Q_\sigma(M) = P$ induces a morphism of sheaves of $\underline{Q}_Y$-modules $\underline{Q}_M|Y \to \underline{Q}_P|Y$, which is isomorphic, as for each $p \in Y$ the map $j_{\sigma,p} : M_p \to Q_\sigma(M)_p$ is an isomorphism. The fact that for any quasicoherent sheaf of $\underline{Q}_Y$-modules on Y we may find a quasicoherent sheaf of $\underline{Q}_R$-modules F on Spec(R) restricting to E on Y, then yields the assertion, since this F is necessarily of the form $\underline{Q}_M$ for some R-module M, which may be chosen σ-closed by the foregoing.

(6.19.) Corollary With assumptions as before, let $j : Y \to X$ denote the canonical inclusion, then for any quasicoherent sheaf of $\underline{Q}_Y$-modules E on Y, the sheaf of $\underline{Q}_X$-modules j_*E on X is quasicoherent.

Proof Let us first check this in the affine case, i.e. assume $X = \text{Spec}(R)$ and $Y = \mathbf{K}(\sigma)$, where σ has finite type in R-mod. Choose $M \in$ R-mod, such that $\underline{Q}_M|Y = E$. Upon replacing M by $Q_\sigma(M)$, we may assume M is σ-closed.

We claim that $j_*E = \underline{Q}_M$. Indeed, let $f \in R$, then on $X(f) \subset \text{Spec}(R)$ we have

$$\Gamma(X(f), j_*E) = \Gamma(X(f) \cap \mathbf{K}(\sigma), \underline{Q}_M|\mathbf{K}(\sigma)) = Q_f(Q_s(M)) = M_f = \Gamma(X(f), \underline{Q}_M),$$

where we have used (5.21.).

To prove the general statement, argue locally as in (4.16.), i.e. choose an open affine covering $\{U_\alpha = \text{Spec}(R_\alpha); \alpha \in A\}$ of X such that $V_\alpha = U_\alpha \cap Y$ is of the form $\mathbf{K}(\sigma_\alpha)$ for some idempotent kernel functor σ_α in R_α-mod and let $i_\alpha : V_\alpha \to U_\alpha$ denote the canonical inclusion. For each $\alpha \in A$ we get $i_*E|U_\alpha = i_{\alpha,*}(E|V_\alpha)$ and from this the result follows by applying a local-global argument and the validity of the statement in the affine case.

Note that we obviously have $i_*E|Y = E$, in the foregoing result!

(6.20.) The classification result (6.18.) also has a global version. Indeed, let X be locally of finite type with respect to a generically closed subset Y. Let $\{U(\alpha) = \text{Spec}(R_\alpha); \alpha \in A\}$ be an open affine covering of X and for each $\alpha \in A$, let $\sigma(\alpha) = \sigma_{U(\alpha) \cap Y}$ be the corresponding idempotent kernel functor in R_α-mod, each of these having finite type, by assumption. If E is an arbitrary quasicoherent sheaf of $\underline{O}_X$-modules on X, then $E|U_\alpha$ is quasicoherent on U_α for each α, hence of the form $\underline{O}_{M(\alpha)}$ for some R_α-module $M(\alpha)$. Assign to any U_α the corresponding sheaf of modules F_α associated to the localization $Q_{\sigma(\alpha)}(M_\alpha)$, then we want to show that these F_α glue together well.

So, let U_1 and U_2 be any two open subsets in the chosen covering and choose some $V = \text{Spec}(R) \subset U_1 \cap U_2$ open and affine (take $V = U_1 \cap U_2$ if X happens to be affine), then we need to verify that the sheaves associated to $Q_{\sigma(1)}(M_1)$ and $Q_{\sigma(2)}(M_2)$ coincide on V.

Assume that V corresponds to some idempotent kernel functor $\tau(i)$ in the category R_i-mod for $i = 1, 2$, then $\tau(i)$ has property (T). Let τ be the idempotent kernel functor in R-mod associated to $V \cap Y \subset V$. Since $E|V = (E|U_i)|V$, we have $Q_{\tau(1)}(M_1) = Q_{\tau(2)}(M_2)$, while $U_i \cap Y \cap V = Y \cap V$ implies that

$$Q_{\tau(i)}(Q_{\sigma(i)}(M_i)) = Q_{\sigma(i)}(Q_{\tau(i)}(M_i)) = Q_\tau(Q_{\tau(i)}(M_i)),$$

whence the assertion.

Now denote by $Q_Y(E)$ the quasicoherent sheaf on X thus obtained. One easily shows that $Q_Y(E)$ is well defined, i.e. independent of the choice of the affine open covering used in its definition. There is a canonical map $j_Y : E \to Q_Y(E)$, defined locally on affines $U = \text{Spec}(R)$ by the morphisms $\Gamma(U, E) \to Q_{Y \cap U}(\Gamma(U, E))$ and passing to the associated quasicoherent sheaves. The kernel of j_Y is denoted by $\sigma_Y E$ (if no ambiguity arises); it may be obtained from the $\sigma_{U \cap Y}\Gamma(U, E)$ by glueing the associated quasicoherent sheaves together. Clearly $\sigma_Y E = E$ if and only if $Q_Y(E) = 0$. We say that E is Y-<u>closed</u> if j_Y is an isomorphism. We now have:

(6.21.) Proposition Let X be a separated scheme, which is locally of finite type with respect to a generically closed subset Y, then there is a one-to-one correspondence between

(6.21.1.) Y-closed quasicoherent sheaves of $\underline{O}_X$-modules on X;

(6.21.2.) quasicoherent sheaves of $\underline{O}_Y = \underline{O}_X|Y$-modules on Y.

Proof This is a trivial consequence of (6.18.), of which this is a global version.

(6.22.) Note The construction of $Q_Y(-)$ in the category $\mathbf{Q}(X, \underline{O}_X)$ of all quasicoherent sheaves of $\underline{O}_X$-modules on X may be viewed as a localization. Indeed, let $\mathbf{T}(Y)$ be the set of all Y-torsion quasicoherent sheaves of $\underline{O}_X$-modules E, i.e. with $\sigma_Y E = E$. One easily sees that E belongs to $\mathbf{T}(Y)$ if and only if $E \in \mathbf{Q}(X, \underline{O}_X)$ and $E|Y = 0$, i.e. we get $\text{Supp}(E) \cap Y = \varnothing$. From this it follows that for any exact sequence of sheaves of $\underline{O}_X$-modules

$$0 \to E' \to E \to E'' \to 0$$

we have $E \in \mathbf{T}(Y)$ if and only if $E' \in \mathbf{T}(Y)$ and $E'' \in \mathbf{T}(Y)$, due to the fact that $\text{Supp}(E) = \text{Supp}(E') \cup \text{Supp}(E'')$. Hence $\mathbf{T}(Y)$ is a localizing subcategory of $\mathbf{Q}(X, \underline{O}_X)$, cf. [Ga].

One easily checks that $\sigma_Y E$ is the largest quasicoherent subsheaf F of E with F in $\mathbf{T}(Y)$. Indeed, obviously $\sigma_Y E$ belongs to $\mathbf{T}(Y)$. On the other hand, if $F \subset E$ and F belongs to $\mathbf{T}(Y)$, then for any open affine $U = \text{Spec}(R)$ inf X, we may find some R-modules M and N with $F|U = \underline{O}_N$, $E|U = \underline{O}_M$ and such that $N \subset M$. By construction $\underline{O}_{N,p} = 0$ for all $p \in Y \cap U = \mathbf{K}(\sigma) \subset \text{Spec}(R)$, hence N is σ-torsion. But then we get $\sigma_Y E|U = \underline{O}_{\sigma M} \supset \underline{O}_N = F|U$ and it follows that $F \subset \sigma_Y E$.

It is fairly straightforward for the reader acquainted with the contents of [Ga] say, to deduce from this that the functor Q_Y may be identified with the localization functor associated to the localizing subcategory $\mathbf{T}(Y)$ of $\mathbf{Q}(X, \underline{O}_X)$. We leave details to the reader. Anyway, at the end of this Section, we will show in an Appendix that the foregoing type of construction may be generalized to the full category $\mathbf{S}(X, \underline{O}_X)$. The reader interested in details may skip the next paragraphs and go to (6.33.) immediately.

(6.23.) Let $(Z, \underline{O}_Z)$ be a ringed space, then we denote by $\mathbf{Q}(Z, \underline{O}_Z)$ the full subcategory of $\underline{\mathbf{S}}(Z, \underline{O}_Z)$ consisting of all the quasicoherent sheaves of $\underline{O}_Z$-modules on Z. We have seen in the foregoing that the restriction map $\mathbf{Q}(X, \underline{O}_X) \to \mathbf{Q}(Y, \underline{O}_Y)$ is surjective, whenever X is separated and Y a generically closed subset of X with respect to which X is locally of finite type. The reader may verify that this result remains valid if we replace the separatedness assumption on X by that of X being locally noetherian (check!).

Let us now briefly consider what happens with coherent sheaves. As usually, we will have to put some rather harmless finiteness assumptions on X or $\underline{O}_X$ in order to make things run smoothly. So, from now on, X will always be assumed to be locally Y-noetherian at least, where Y is a fixed generically closed subset of X. If we assume moreover X to be quasicompact, i.e. X is Y-noetherian, then clearly $\underline{O}_Y = \underline{O}_X|Y$ is a noetherian object in $\mathbf{Q}(Y, \underline{O}_Y)$. Indeed, one easily reduces to the affine case, since X may be covered by a finite number of open affines. In this case, let $X = \mathrm{Spec}(R)$ and $Y = \mathbf{K}(\sigma)$ for some idempotent kernel functor σ such that R is σ-noetherian. Since there is a one-to-one correspondence between the objects in $\mathbf{Q}(Y, \underline{O}_Y)$ and σ-closed R-modules in this case, the fact that $Q_\sigma(R)$ is a noetherian object in (R, σ)-mod proves the assertion.

Moreover, the same reasoning shows that in this case sheaves in $Q(Y, \underline{O}_Y)$ which are locally of finite type are also noetherian objects.

Let us call a quasicoherent sheaf E of $\underline{O}_X$-modules Y-<u>coherent</u> if and only if for all open affine U = Spec(R) in X, the module $\Gamma(U, E)$ is $\sigma_{U \cap Y}$-finitely generated over R - or, equivalently, if $\Gamma(U, E)$ is $\sigma_{U \cap Y}$-noetherian, since the ring R is $\sigma_{U \cap Y}$-noetherian by assumption.

(6.24.) Proposition Let E be a quasicoherent sheaf of $\underline{O}_X$-modules on X, then E is Y-coherent if and only if there exists a covering of X by open affines of the form U = Spec(R), such that for each of these the R-module $\Gamma(U, E)$ is $\sigma_{U \cap Y}$-finitely generated.

Proof Only one implication needs a proof, so let U = Spec(R) be an open affine subset of X and assume that X may be covered by open affines U_i, with the above property. It follows that U possesses a basis of open affines V = Spec(A) such that $\Gamma(V, E)$ is $\sigma_{V \cap Y}$-finitely generated overA.

This permits us to reduce the problem to the following : if there exist elements f(1), ... , f(n) in R which generate the unit ideal, such that, denoting $\sigma_{U \cap Y}$ by σ, the rings R and $R_{f(i)}$ are σ-noetherian and such that for each of the indices i the localization $M_{f(i)}$ of an R-module M is σ-finitely generated (and hence σ-noetherian), then is it true that M is σ-finitely generated? This may be checked exactly as in (6.16.). We leave details to the reader.

(6.25.) Lemma Our assumptions imply that $\underline{O}_Y = \underline{O}_X|Y$ is coherent.

Proof We have to check the following : if U is an open subset of Y, then for any morphism $\phi : (\underline{O}_Y)^p|U \to \underline{O}_Y|U$ of sheaves of $\underline{O}_U$-modules, the kernel Ker(ϕ) of this map is locally of finite type. One reduces easily to the affine case, i.e. we may assume X = Spec(R), where R is a σ-noetherian ring and Y = $K(\sigma)$. The open subset U of Y then corresponds to some idempotent kernel functor $\tau \geq \sigma$ in R-mod. Taking global sections yields a map $\psi = \Gamma(U, \phi) : Q_\tau(R)^p \to Q_\tau(R)$, say with kernel K. Passing to the associated sheaves on X and restricting to U yields that $\mathrm{Ker}(\phi) = \underline{Q}_K|U$, hence $\mathrm{Ker}(\phi)$ is quasicoherent. Now, $K \subset Q_\tau(R)^p$ implies that K is τ-finitely generated, as R is also τ-noetherian, hence $\underline{Q}_K|U$ is of finite type indeed, proving that $\underline{Q}_Y$ is coherent.

(6.26.) Proposition Let E be a quasicoherent sheaf of $\underline{Q}_X$-modules on X, then E is Y-coherent if and only if E|Y is a coherent sheaf of $\underline{Q}_Y$-modules on Y

Proof Let E be Y-coherent, then we want to show that E|Y is coherent on Y. It clearly suffices to verify this in the affine case, so, with notations as in the foregoing Lemma, we may assume that $E = \underline{Q}_M$, where M is σ-finitely generated, hence σ-finitely presented, i.e. there exists some R-linear map $u : N \to M$, where N is finitely presented and where $\mathrm{Ker}(u)$ and $\mathrm{Coker}(u)$ are σ-torsion. It follows that $\underline{Q}_N$ is finitely presented over $\underline{Q}_R$, so $\underline{Q}_N|Y$ is a finitely presented sheaf of $\underline{Q}_Y$-modules. Moreover, our choice of u shows that $\underline{Q}_N|Y$ and $\underline{Q}_M|Y$ are isomorphic, so E is finitely presented, hence coherent over $\underline{Q}_Y$, as $\underline{Q}_Y$ is a coherent sheaf of rings on Y.

Conversely, assume E is a quasicoherent sheaf of $\underline{Q}_X$-modules such that E|Y is coherent on Y, then we have to show that E is Y-coherent. Let U be an open affine subset of X, then $E|U \cap Y$ is coherent on $U \cap Y$, so we reduce the problem to the case where $X = \mathrm{Spec}(R)$ is affine and $Y = \mathbf{K}(\sigma) \subset \mathrm{Spec}(R)$. We have an R-module M such that $\underline{Q}_M|Y$ is $\underline{Q}_Y$-coherent.

Using the fact that both the property of being coherent and that of being Y-coherent is local on the base space, we reduce to the question :

if $\underline{Q}_Y^p \to \underline{Q}_Y^q \to \underline{Q}_M|Y \to 0$ is a presentation of $\underline{Q}_M|Y$ for some positive integers p, q, does it follow that M is σ-finitely generated.

Let N denote the cokernel of the map $Q_\sigma(R)^p \to Q_\sigma(R)^q$ induced by taking global sections over Y. We obviously have $\underline{Q}_N|Y = \underline{Q}_M|Y$ as we have pointed out before, so it suffices to verify that N is σ-finitely generated over R. Now, the canonical map $Q_\sigma(R)^q \to N$ yields a morphism $R^q \to N$ with σ-torsion kernel and cokernel, which proves the assertion.

(6.27.) Lemma Let X be locally noetherian and Y a generically closed subset of X, then for any quasicoherent sheaf of $\underline{O}_X$-modules F on X and any coherent sheaf of $\underline{O}_Y$-modules $G \subset F|Y$ on Y, there exists a coherent sheaf of $\underline{O}_X$-modules $G' \subset F$ on X such that $G'|Y = G$.

Proof This result and its proof should be compared to (I.6.9.6.) in [GD]. Assume $\{U_\alpha; \alpha \in A\}$ is a covering of X by open affines and assume A to be well-ordered. We put $Y_\alpha = Y \cup V_\alpha$, where $V_\alpha = \cup\{U_\beta; \beta < \alpha\}$, then X is the increasing union of the Y_α. We will construct by induction a family $(G'_\alpha; \alpha \in A)$, where $G'_\alpha \subset F|Y_\alpha$ is coherent on Y_α, where $G'_\alpha|V_\beta = G'_\beta$ for $\beta < \alpha$ and where $G'_\alpha|Y = G$. It then follows that there exists a unique G' on X with $G'|U_\alpha = G'_\alpha$ and this clearly satisfies the requirements.

So assume the G'_β defined for all $\beta < \alpha$. If α is a limit ordinal, let G'_α be the unique $G'_\alpha \subset F|Y_\alpha$ such that $G'_\alpha|Y_\beta = G'_\beta$ for all $\beta < \alpha$; this makes sense, since the U_β with $\beta < \alpha$ cover U_α.

Otherwise, i.e. if α is not a limit ordinal, it has a predecessor β. Then $Y_\alpha = Y_\beta \cup U_\beta$ and it will be sufficient to define some coherent $G''_\beta \subset F|U_\beta$ with the property that

$$G''_\beta|Y_\beta \cap U_\beta = G'_\beta|Y_\beta \cap U_\beta,$$

because then one may take $G'_\alpha \subset F|Y_\alpha$ to be the unique (coherent) sheaf with $G'_\alpha|Y_\beta = G'_\beta$ and $G'_\alpha|U_\beta = G''_\beta$.

Now, U_β is affine, so it has the form $U_\beta = \mathrm{Spec}(R)$ for some noetherian ring R, and $Y_\beta \cap U_\beta \subset U_\beta$ is generically closed, say of the form $\mathbf{K}(\sigma)$, where σ has finite type in R-mod (since R is noetherian!). As G'_β is coherent on Y_β, it follows that $G'_\beta|Y_\beta \cap U_\beta$ is coherent on $Y_\beta \cap U_\beta$. Now, by (6.17.) we may find some quasicoherent E on U_β, with $E|Y_\beta \cap U_\beta =$

$G'_\beta | Y_\beta \cap U_\beta$, since σ has finite type and by (6.26.), it follows that E is then $Y_\beta \cap U_\beta$-coherent. By definition, the module $M = \Gamma(U_\beta, E)$ is σ-finitely generated over R, i.e. we may find $N \subset M$ finitely generated, such that M/N is σ-torsion. But then $G''_\beta = \underline{O}_N$ is certainly coherent, since R is noetherian and $G''_\beta | Y_\beta \cap U_\beta = \underline{O}_M | Y_\beta \cap U_\beta = G'_\beta | Y_\beta \cap U_\beta$, since $\underline{O}_M = E$, the scheme being affine. This proves the assertion.

For any ringed space $(Z, \underline{O}_Z)$, denote by $\mathbf{C}(Z, \underline{O}_Z)$ the full subcategory of $\underline{S}(Z, \underline{O}_Z)$, consisting of coherent sheaves of $\underline{O}_Z$-modules on Z. We then have:

(6.28.) Corollary For any separated locally noetherian scheme X and any generically closed subset Y of X, the restriction map

$$\mathbf{C}(X, \underline{O}_X) \to \mathbf{C}(Y, \underline{O}_Y)$$

is surjective.

Proof If E is a quasicoherent sheaf of $\underline{O}_Y$-modules on Y, then the sheaf $F = i_* E$, with $i : Y \to X$ denoting the canonical inclusion, is quasicoherent on X. Note also that $F|Y = E$, so we may apply the foregoing Lemma, in case E is coherent, by taking $G = F|Y = E$.

Note If one takes into account the remarks made at the beginning of (6.23.), then it follows that the separatedness assumption may be dropped in the foregoing result. We leave this verification as an exercise to the reader.

The foregoing also permits a partial generalization of (5.24.), as well as a stronger version of (I.6.9.17.) in [GD]. Let Y be a subset of X, then we write $\underline{L}(Y)$ for the set of all quasicoherent sheaves of ideals I of $\underline{O}_X$ with the property that $\text{Supp}(\underline{O}_X/I) \cap Y = \varnothing$. Here, as usually, for any sheaf of abelian groups E, we denote by $\text{Supp}(E)$ the set of all points $x \in X$ with $E_x \neq 0$. The set $\underline{L}(Y)$ is ordered by inclusion and behaves formally as a Gabriel filter. We then have:

(6.29.) Proposition Let X be a noetherian scheme and let Y be a generically closed subset of X. If E is a coherent sheaf of $\underline{O}_Y$-modules on Y and F a quasicoherent sheaf of $\underline{O}_X$-modules on X, then the set $[E, F|Y]$ may be described by

$$[E, F|Y] = \lim [IE', F],$$

where E' is a coherent sheaf of $\underline{O}_X$-modules such that $E'|Y = E$ and where I runs through $\underline{L}(Y)$.

Proof Of course $[E, F|Y]$ denotes the set of all morphisms $E \to F|Y$ in $\underline{S}(Y, \underline{O}_Y)$ and $[IE', F]$ that of all $IE' \to F$ in $\underline{S}(X, \underline{O}_X)$. The extension E' of E exists by (6.28.), if one takes into account the note following its proof. Otherwise, assume X to be separated.

Let us first look at the affine case. So, let R be a noetherian ring and let $Y = K(\sigma)$ for some idempotent kernel functor σ in R-mod. In this case any J im $\underline{L}(Y)$ is of the form $\underline{O}_I$ for some ideal I of R and for any p in $\text{Spec}(R)$, we have $p \in \text{Supp}(\underline{O}_R/\underline{O}_I)$ if and only if $I_p \neq R_p$. Hence we obtain $\underline{O}_I \in \underline{L}(Y)$ if and only if $I_p = R_p$ for all $p \in Y$, i.e. if $I \in L(\sigma) = L(Y)$. Moreover, in the present situation $F = \underline{O}_N$ and $E' = \underline{O}_M$ for some R-modules M and N, with M finitely generated. It follows that $[IE', F] = [IM, N]$.

Now , since M is finitely generated, hence finitely presented (as R is noetherian) we may find an exact sequence of the form

$$R^p \to R^q \to M \to 0,$$

for some positive integers p and q. Let $T = \mathrm{Im}(R^p \to R^q) \subset R^q$ and for any J in $\mathbf{L}(\sigma)$, let M_J denote the cokernel of the induced map $JT \to JR^q$. Since R is noetherian, we may apply Krull's Theorem (cf. [AM]), which yields a positive integer n such that $T \cap J^nR^q \subset JT$.

Consider the exact commutative diagram

$$0 \to T \cap J^nR^q \to J^nR^q \to J^nM \to 0$$
$$\downarrow \qquad\qquad \downarrow \qquad\quad \downarrow$$
$$0 \to \quad JT \longrightarrow JR^q \to M_J \to 0$$

where the map $J^nR^q \to J^nM$ in the top row is induced by the map $R^q \to M$ above. Since $JT \cap J^nR^q \subset J^nR^q$, clearly $J^nM \to M_J$ is injective. The exact sequences

$$JR^p \to JR^q \to M_J \to 0$$

$(J \in \mathbf{L}(\sigma))$ yield an exact sequence

$$0 \to \lim\, [M_J, N] \to \lim\, [JR^q, M] \to \lim\, [JR^p, M],$$

where each time the limit is taken over all $J \in \mathbf{L}(\sigma)$. On the other hand, the exact commutative diagrams above yield an exact commutative diagram

$$0 \to \lim\, [M_J, N] \;\to\; \lim\, [J^nR^q, N] \to \lim\, [JT, N]$$
$$\alpha \downarrow \qquad\qquad \beta \downarrow \qquad\qquad \gamma \downarrow$$
$$0 \to \lim\, [J^nM, N] \to \lim\, [JR^q, N] \to \lim\, [T \cap J^nR^q, N],$$

where $J \in \mathbf{L}(\sigma)$. Obviously β is an isomorphism. We claim that α is also an isomorphism and for this to hold, it suffices to check that γ is an isomorphism. But, for any $J \in \mathbf{L}(\sigma)$, we may find positive integers n, m (depending on J) such that

$$JT \supset T \cap J^nR \supset J^nT \supset T \cap J^mR.$$

This yields maps

$$\lim [JT, N] \to \lim [T \cap J^nR^q, N] \to \lim [J^nT, N] \to \lim [T \cap J^mR, N]$$

and with $\lim [T \cap J^nR^q, N] = \lim [T \cap J^mR^q, N]$ resp. $\lim [JT, N] = \lim [J^nT, N]$, one easily checks that the composition of any two of consecutive ones of these maps yields the identity, hence that α is an isomorphism, indeed. We thus find an exact sequence

$$0 \to \lim [JM, N] \to \lim [JR^q, N] \to \lim [JR^p, N].$$

On the other hand, the exact sequence $R^p \to R^q \to M \to 0$ also yields an exact sequence of quasicoherent sheaves of $\underline{O}_R|\mathbf{K}(\sigma)$-modules on $\mathbf{K}(\sigma)$

$$0 \to [\underline{O}_M|Y, \underline{O}_N|Y] \to [\underline{O}_{R^q}|Y, \underline{O}_N|Y] \to [\underline{O}_{R^p}|Y, \underline{O}_N|Y]$$

One easily verifies that these fit into an exact commutative diagram

$$
\begin{array}{ccccccc}
0 & \to & \lim [JM, N] & \to & \lim [JR^q, N] & \to & \lim [JR^p, N] \\
 & & \downarrow & & \downarrow & & \downarrow \\
0 & \to & [\underline{O}_M|Y, \underline{O}_N|Y] & \to & [\underline{O}_{R^q}|Y, \underline{O}_N|Y] & \to & [\underline{O}_{R^p}|Y, \underline{O}_N|Y]
\end{array}
$$

To conclude (in the affine case), it suffices to check that the latter vertical maps are isomorphisms. But for this, it is clearly also sufficient to verify that $\lim [J, N] = [\underline{O}_R|Y, \underline{O}_N|Y]$. Now, $\lim [J, N] = Q_\sigma(N) = \Gamma(\mathbf{K}(\sigma), \underline{O}_N) =$

$[\underline{Q}_R|Y, \underline{Q}_N|Y]$, where we used the fact that σ is stable, as R is noetherian, cf. (5.20.).

For the general case, i.e. X not necessarily affine, cover X by a finite number of open affine, whose pairwise intersections are also covered by a finite number of affines (they are affine themselves, if X is separated!), then an easy local global argument finishes the proof.

(6.30.) Corollary Let R be a noetherian ring and σ an idempotent kernel functor in R-mod, then $\Gamma(\mathbf{K}(\sigma), \underline{Q}_M) = Q_\sigma(M)$ for any R-module M.

Proof Take $E' = \underline{Q}_R$ and $F = \underline{Q}_M$ in (6.29.). Note that (5.24.) is actually stronger than the present result!

(6.31.) Recall that a <u>closed subscheme</u> of a scheme $(X, \underline{O}_X)$ is a scheme $(Y, \underline{O}_Y)$ together with a scheme morphism $(i, i^\#) : (Y, \underline{O}_Y) \to (X, \underline{O}_X)$ such that Y is a closed subset of X and $i : Y \to X$ is the canonical inclusion, and such that the morphism $i^\# : \underline{O}_X \to i_*\underline{O}_Y$ on sheaves is surjective, i.e. an epimorphism of sheaves of abelian groups. The ideal sheaf $\underline{J}_Y$ of Y is then defined to be the kernel of $i^\#$. It is a quasicoherent sheaf of ideals of $\underline{O}_X$, which is coherent if X is noetherian. Actually, every quasicoherent sheaf of $\underline{O}_X$-ideals on X accurs in this way. Indeed, let $\underline{J}$ be such a sheaf and let $Y = \mathrm{Supp}(\underline{O}_X/\underline{J})$, then one may check that $(Y, \underline{O}_X/\underline{J})$ is a closed subscheme of $(X, \underline{O}_X)$ and that $\underline{J} = \underline{J}_Y$.

(6.32.) Corollary ([GD], I.6.9.17.) Let X be a noetherian scheme, $\underline{J}$ a quasicoherent sheaf of ideals of $\underline{O}_X$ and V the closed subscheme of X defined by $\underline{J}$. Let $Y = X - V$ and assume E, E' and F are as in (6.29.), then there is an isomorphism

$$\lim [\underline{J}^n E', F] = [E, F|Y].$$

Proof The set of all powers $\underline{J}^n$ forms a cofinal subset of $\underline{L}(Y)$, as one easily checks by looking at a finite open covering. Indeed, one then reduces the problem to the affine case, i.e. $X = \mathrm{Spec}(R)$ for some noetherian ring R and $\underline{J} = \underline{Q}_I$. If $\underline{Q}_J \in \underline{L}(Y)$, then $J_p = R_p$ for all $p \in Y = \mathbf{K}(\sigma_I) = X(I)$, i.e. $J \in \mathbf{L}(I)$ and $I^n \subset J$ for some positive integer n. Hence $\underline{J}^n = \underline{Q}_{I^n} \subset \underline{Q}_J$, indeed. The result then follows immediately from (6.29.)

(6.33.) Appendix : The local cohomology point of view.

We will conclude this Section by briefly indicating the relation between the foregoing constructions and local cohomology functors. In this part we assume some familiarity with the notion of derived functors. Our set-up is inspired by [Su]. The reader is also invited to take a look at related results in [Ha2].

Let $(X, \underline{O}_X)$ denote a ringed space and Z a subset of X. We denote by $Y = X - Z$ the complement of Z and for convenience's sake we will always assume Y to be generically closed, hence (equivalently) that Z is closed under specialization. Anyway, the reader may easily check that our constructions only depend upon the generic closure $Y^\wedge$ of Y.

Denote by $\underline{S}_Z(X, \underline{O}_X)$ the full subcategory of $\underline{S}(X, \underline{O}_X)$ consisting of all sheaves of $\underline{O}_X$-modules E with $\text{Supp}(E) \subset Z$. If X is a scheme, which is locally of finite type with respect to Y, then we have $\mathbf{T}(Y) \subset \underline{S}_Z(X, \underline{O}_X)$. More precisely, $\mathbf{T}(Y) = \underline{S}_Z(X, \underline{O}_X) \cap \mathbf{Q}(X, \underline{O}_X)$.

Since for any exact sequence

$$0 \to E' \to E \to E'' \to 0$$

in $\underline{S}(X, \underline{O}_X)$ we have $\text{Supp}(E) = \text{Supp}(E') \cup \text{Supp}(E'')$, it follows as before that E belongs to $\underline{S}_Z(X, \underline{O}_X)$ if and only if E' and E'' belong to $\underline{S}_Z(X, \underline{O}_X)$. As a consequence, $\underline{S}_Z(X, \underline{O}_X)$ is an abelian subcategory of $\underline{S}(X, \underline{O}_X)$ and the canonical embedding $i_Z : \underline{S}_Z(X, \underline{O}_X) \to \underline{S}(X, \underline{O}_X)$ is exact. The embedding i_Z possesses a right adjoint

$$\Gamma_Z : \underline{S}(X, \underline{O}_X) \to \underline{S}_Z(X, \underline{O}_X),$$

sending any sheaf of $\underline{O}_X$-modules E on X to the largest subsheaf F in $\underline{S}(X, \underline{O}_X)$ with $F \in \underline{S}_Z(X, \underline{O}_X)$. Such an F actually exists and is easily seen to be given by

$$U \rightarrow \Gamma_{Z \cap U}(U, E|U),$$

where $\Gamma_{Z \cap U}(U, E|U)$ consists of all sections $s \in \Gamma(U, E)$ with $s_x = 0$ for all x in $Y \cap U$, i.e. with $\mathrm{Supp}(\underline{O}_U f) \subset Z \cap U$.

(6.34.) Lemma If X is a scheme, which is locally of finite type with respect to a generically closed subset Y and if $Z = X - Y$, then for any E in $Q(X, \underline{O}_X)$, we have $\Gamma_Z E = \sigma_Y E$.

Proof We may obviously reduce the problem to the affine case, i.e. we may take $Y = \mathbf{K}(\sigma) \subset \mathrm{Spec}(R) = X$ for some R-module M. For any $U = X(f)$ in $\mathrm{Spec}(R)$, we have $s \in (\Gamma_Z E)(U)$ if and only if $s \in \Gamma(X(f), \underline{O}_M|X(f))$ and $s_p = 0$ for all p in $X(f) \cap Y$. But $\Gamma(X(f), \underline{O}_M) = M_f$ and the set of all $s \in M_f$ with $s_p = 0$ for all $p \in X(f) \cap \mathbf{K}(\sigma)$ is exactly $\sigma M_f = (\sigma M)_f$. Hence s belongs to $(\sigma_Y E)(U)$ and conversely.

(6.35.) Note An alternative proof may be given as follows. First note that if E is a sheaf (of abelian groups) on X and G a subsheaf of E|Y on Y, then $i_* G \subset i_*(E|Y)$. There is a canonical morphism $\rho_E : E \rightarrow i_*(E|Y)$ and we let $G' = (\rho_E)^{-1}(G) \subset E$. Obviously, for any open subset U of X the set $\Gamma(U, G')$ consists of all $s \in \Gamma(U, E)$ such that $s|U \cap Y \in \Gamma(U \cap Y, G)$, where $s|Y \cap U$ is the image of s through

$$\Gamma(U, E) \rightarrow \lim \Gamma(W, E) \rightarrow (E|Y)(Y \cap U),$$

where W runs through the open subsets of X containing $Y \cap U$. It thus follows that $G'|Y = G$ and G' is the largest subsheaf of E with this property.

Now, if we work in the set-up of the Lemma, if H is a quasicoherent sheaf on Y, then i_*H is a quasicoherent sheaf on X. So, let E be quasicoherent on X and G quasicoherent on Y, with $G \subset E|Y$, then it easily follows that G' is quasicoherent. We have pointed out in (6.22.) that $\sigma_Y E$ is the largest quasicoherent sheaf $F \subset E$ with the property that $F_y = 0$ for all $y \in Y$, i.e. with $F|Y = 0$, so applying the foregoing to $G = 0 \subset E|Y$, we find that $\Gamma_Z E = \sigma_Y E$, indeed.

(6.36.) We denote the right derived functors of Γ_Z by H^i_Z ($i \geq 0$). For example, to calculate $H^1_Z(E)$ for some $E \in \underline{S}(X, \underline{O}_X)$, let I be an injective sheaf in $\underline{S}(X, \underline{O}_X)$ containing E, then there is an exact sequence

$$0 \to \Gamma_Z E \to \Gamma_Z I \to \Gamma_Z(I/E) \to H^1_Z(E) \to 0.$$

For $i = 0$, we of course have $H^0_Z = \Gamma_Z$.
More generally, let $Z' \subset Z$, then as in [Su] we write $\Gamma_{Z/Z'} E = \Gamma_Z E / \Gamma_{Z'} E$ and $H^i_{Z/Z'} = R^i \Gamma_{Z/Z'}$. In particular, we then have $H^i_{Z/\emptyset} = H^i_Z$ ($i \geq 0$).
For any $Z'' \subset Z' \subset Z$ we have an exact sequence of functors

$$0 \to \Gamma_{Z'/Z''} \to \Gamma_{Z/Z''} \to \Gamma_{Z/Z'} \to 0,$$

from which one deduces a long exact cohomology sequence, which, for the choice $\emptyset \subset Z \subset X$, yields as a special case, for any $E \in \underline{S}(X, \underline{O}_X)$ an exact sequence

$$0 \to \Gamma_Z E \to E \to H^0_{X/Z} E \to H^1_Z E \to 0.$$

Here both $H^1_Z E$ and $\Gamma_Z E$ belong to $\underline{S}_Z(X, \underline{O}_X)$, hence $j_E : E \to H^0_{X/Z} E$ is a Z-isomorphism. More generally, we say that a morphism $u : E \to F$ in $\underline{S}(X, \underline{O}_X)$ is a Z-isomorphism if and only if the induced maps $u_x : E_x \to F_x$

are isomorphic for all $x \in X - Z$ or equivalently, if $\mathrm{Ker}(u)$ and $\mathrm{Coker}(u)$ belong to $\underline{S}_Z(X, \underline{O}_X)$.

(6.37.) Proposition For any $E \in \underline{S}(X, \underline{O}_X)$ the following assertions are equivalent:

(6.37.1.) $j_E : E \to H^0{}_{X/Z}E$ is a monomorphism (resp. an isomorphism);

(6.37.2.) $\Gamma_Z E = 0$ (resp. and $H^1{}_Z(E) = 0$);

(6.37.3.) for any Z-isomorphism $u : M \to N$ in $\underline{S}(X, \underline{O}_X)$, the induced map

$$[N, E] \to [M, E]$$

is injective (resp. bijective).

Proof The equivalence of (1) and (2) is trivial. So, assume (1) or (2) and let us prove (3). First note that $[T, E] = 0$ for all $T \in \underline{S}_Z(X, \underline{O}_X)$. Indeed, any $f : T \to E$ factorizes (by functoriality) through $\Gamma_Z E = 0$, showing that $f = 0$, indeed. Now, assume that u is epimorphic, then it fits into an exact sequence

$$0 \to T \to M \to N \to 0$$

in $\underline{S}(X, \underline{O}_X)$, with $T \in \underline{S}_Z(X, \underline{O}_X)$. But, then we obtain an exact sequence

$$0 \to [N, E] \to [M, E] \to [T, E],$$

hence $[N, E] = [M, E]$. We thus have reduced the proof to the case where u is monomorphic.

So, assume that we have an exact sequence

$$0 \to M \to N \to K \to 0,$$

where $K \in \underline{S}_Z(X, \underline{O}_X)$, then we get an exact sequence

$$0 \to [K, E] \to [N, E] \to [M, E],$$

hence $[N, E] \subset [M, E]$, since $[K, E] = 0$ by the same argument as before. Assume on the other hand, that we also have $H^1_Z E = 0$, then $H^0_{X/Z} E = E$. Choose any $f \in [M, E]$, then f extends to a morphism $f' : H^0_{X/Z} M \to H^0_{X/Z} E = E$ and since u induces an isomorphism $v = H^0_{X/Z} M \to H^0_{X/Z} N$, with inverse w, we obtain a morphism $g = f'wj_N : N \to E$, with $gu = f$, proving that the map $[N, E] \to [M, E]$ is also surjective.

Finally, to prove (3) $\Rightarrow$ (2), consider the trivial Z-isomorphism $0 \to \Gamma_Z E$, which yields $[\Gamma_Z E, E] = 0$, hence $\Gamma_Z E = 0$. If the map $[H^0_{X/Z} E, E] \to [E, E]$ is surjective, then the exact sequence

$$0 \to E \to H^0_{X/Z} E \to H^1_Z E \to 0$$

splits. But then $H^0_{X/Z} E$ would have a nontrivial direct factor with support in Z, which is impossible unless $H^1_Z E = 0$

Note that in the proof of the foregoing result, we have used several times the following result, which is an almost trivial consequence of the definitions.

(6.38.) Lemma For any $E \in \underline{S}(X, \underline{O}_X)$, we have $\Gamma_Z H^0_{X/Z} E = 0$.
If $E \in \underline{S}_Z(X, \underline{O}_X)$, then $H^0_{X/Z} E = H^1_Z E = 0$.
Proof It clearly suffices to verify the first statement for $E \in \underline{S}(X, \underline{O}_X)$, which is injective. But in this case $j_E : E \to H^0_{X/Z} E$ is an epimorphism, hence with F defined by $F = (j_E)^{-1}(\Gamma_Z H^0_{X/Z} E)$, we have an exact sequence

$$0 \to \Gamma_Z F \to F \to \Gamma_Z H^0_{X/Z} E.$$

This proves that $F \in \underline{S}_Z(X, \underline{O}_X)$, so $F = \Gamma_Z E$ and so $\Gamma_Z H^0_{X/Z} E = 0$.

The other assertion is even easier : if $E \in \underline{S}_Z(X, \underline{O}_X)$, then $\Gamma_Z E \to E$ is an isomorphism, hence $H^0_{X/Z} E = H^1_Z E$, and from $H^1_Z E \in \underline{S}_Z(X, \underline{O}_X)$, it follows that $H^0_{X/Z} E = \Gamma_Z H^0_{X/Z} E = 0$.

(6.39.) An $\underline{O}_X$-module E on X is said to be Y-<u>pure</u> resp. Y-<u>closed</u> if the conditions of (6.37.) hold. Note that in [Su] one speaks of a Z-pure resp. Z-closed $\underline{O}_X$-module. It follows easily from the foregoing Lemma that the sheaf $H^0_{X/Z} E$ is Y-pure for any $E \in \underline{S}(X, \underline{O}_X)$. Moreover, if E is Y-pure, then $H^0_{X/Z} E$ is Y-closed, as follows easily from the foregoing. It is thus clear that $H^0_{X/Z} H^0_{X/Z} E$ is Y-closed for any E belonging to $\underline{S}(X, \underline{O}_X)$ and that, by composing the two "j-maps", there is a morphism

$$\mu_{X/Z} : E \to Cl_{X/Z} E = H^0_{X/Z} H^0_{X/Z} E.$$

We call $Cl_{X/Z} E$ (endowed with this morphism $\mu_{X/Z}$) the Y-<u>closure</u> of E. Again [Su] calls this the Z-closure of E.

We leave it as an exercise to the reader to verify that the full subcategory $\underline{S}_{X/Z}(X, \underline{O}_X)$ of $\underline{S}(X, \underline{O}_X)$ of all Y-closed sheaves of $\underline{O}_X$-modules on X is a strict Giraud subcategory of $\underline{S}(X, \underline{O}_X)$, with reflector

$$Cl_{X/Z} : \underline{S}_{X/Z}(X, \underline{O}_X) \to \underline{S}_{X/Z}(X, \underline{O}_X).$$

The next result explains why we call $Cl_{X/Z} E$ the Y-closure of E.

(6.40.) Proposition Let X be a locally noetherian scheme and Y a generically closed subset of X. For any quasicoherent sheaf of $\underline{O}_X$-modules E on X we then have

$$Cl_{X/Z}(E) = Q_Y(E).$$

Proof Although the locally noetherian hypothesis is not strictly necessary here, we have preferred to include it for simplicity's sake. Indeed, it implies all quasicoherent sheaves associated to an injective module to be flabby, cf. (6.14.).

It is clear that it suffices to prove the result in the affine situation. So, assume that $X = \mathrm{Spec}(R)$ for some noetherian ring R and $Y = \mathbf{K}(\sigma)$ for some idempotent kernel functor σ in R-mod. The sheaf E is then of the form $\underline{Q}_M$ for some R-module M. If we put $N = Q_\sigma(M)$, then we wish to show that $Cl_{X/Z}(\underline{Q}_M) = \underline{Q}_N$. Now, there is an exact sequence

$$0 \to \sigma M \to M \to Q_\sigma(M) \to T \to 0,$$

with T being σ-torsion. This sequence induces an exact sequence of sheaves

$$0 \to \underline{Q}_{\sigma M} \to \underline{Q}_M \to \underline{Q}_N \to \underline{Q}_T \to 0.$$

Here $\underline{Q}_{\sigma M} = \sigma_Y \underline{Q}_M = \Gamma_Z \underline{Q}_M$ and $\Gamma_Z \underline{Q}_T = \sigma_Y \underline{Q}_T = \underline{Q}_{\sigma T} = \underline{Q}_T$, so we derive an isomorphism $Cl_{X/Z}\underline{Q}_M = Cl_{X/Z}\underline{Q}_N$. We want to show that $\underline{Q}_N$ is Y-closed.

We need some preparations first. The derived functor $R^1\sigma$ may be calculated using injectives: if M is an R-module and E its injective hull, then there is an exact sequence

$$(\text{*}) \qquad 0 \to \sigma M \to \sigma(E/M) \to (R^1\sigma)(M) \to 0.$$

In particular, if M is σ-closed, then it is σ-torsionfree, hence so is E and it follows that E is σ-closed as well. From this it follows that the module E/M is σ-torsionfree, hence $(R^1\sigma)(M) = 0$.

Now, again in the general case, i.e. M not necessartily σ-closed, we have an exact sequence of sheaves

$$(\text{**}) \qquad 0 \to \Gamma_Z \underline{O}_M \to \Gamma_Z \underline{O}_E \to \Gamma_Z \underline{O}_{E/M} \to H^1{}_Z \underline{O}_M \to H^1{}_Z \underline{O}_E$$

and we will see below that $H^1{}_Z \underline{O}_E = 0$. The exact sequence (*) yields an exact sequence of quasicoherent sheaves, and comparing this with (**), we find that $H^1{}_Z \underline{O}_M = \underline{O}_P$, where $P = (R^1\sigma)(M)$. In particular, it follows that if M is σ-closed, then $H^1{}_Z \underline{O}_M = 0$. Since in this case we also have $\Gamma_Z \underline{O}_M = \underline{O}_{\sigma M} = 0$, we thus find that any σ-closed R-module M yields a Y-closed quasicoherent sheaf $\underline{O}_M$ on X, as we asserted.

It thus remains to verify that $H^1{}_Z \underline{O}_E = 0$, when E is an injective R-module. Pick some injective sheaf K in $\underline{S}(X, \underline{O}_X)$, fitting in an exact sequence

$$0 \to \underline{O}_E \to K \to L \to 0.$$

Restricting to Y and taking sections, we find an exact sequence

$$0 \to \Gamma(Y, \underline{O}_E) \to \Gamma(Y, K) \to \Gamma(Y, L).$$

Since E is injective, the "restriction map" $\Gamma(X, \underline{O}_E) \to \Gamma(Y, \underline{O}_E)$, which reduces to $E \to Q_\sigma(E)$, is surjective, cf. (6.14.). Denote this by $s \to s|Y$. Assume that s belongs to $(\Gamma_Z L)(X)$, i.e. $s \in \Gamma(X, L)$ and $s_y = 0$ for all $y \in Y$, then we may find some $t \in \Gamma(X, K)$ with $p(X)(t) = s$ by (2.29.), since $\underline{O}_E$ is flabby. Clearly $p(Y)(t|Y) = s|Y = 0$, so $t|Y = i(Y)(a)$ for some $a \in \Gamma(Y, \underline{O}_E)$. Let $b \in \Gamma(X, \underline{O}E) = E$ be such that $b|Y = a$, then we have $p(X)(t - b) = p(X)(t) = s$ and $(t - b)_y = t_y - b_y = t_y - a_y = 0$. This proves that the sequence

$$0 \to \Gamma_Z \underline{O}_E \to \Gamma_Z K \to \Gamma_Z L \to 0$$

is exact, hence, since $H^1{}_Z K = 0$, we find that $H^1{}_Z \underline{O}_E = 0$, indeed. This finishes the proof.

7 GLOBAL RELATIVE INVARIANTS

(7.1.) Throughout $(X, \underline{O}_X)$ denotes a ringed space and for any pair of sheaves of $\underline{O}_X$-modules E and F, we will write **Hom** (E, F) resp. $E \otimes F$ without mentioning $\underline{O}_X$, whenever morphisms or tensorproducts are taken over $\underline{O}_X$.

We have seen before that the Picard group $\text{Pic}(X, \underline{O}_X)$ or $\text{Pic}(X)$ is the set of all isomorphism classes [E] of invertible sheaves of $\underline{O}_X$-modules E on X, i.e. which are locally free of rank one. For such a sheaf, $E^* = \textbf{Hom}(E, \underline{O}_X)$ is invertible too and the canonical pairing induces an isomorphism $E \otimes F = \underline{O}_X$. From this one easily deduces that $\text{Pic}(X)$ may be endowed with a group structure, putting $[E] \cdot [F] = [E \otimes F]$ and with $[E]^{-1} = [E^*]$.

Of course, any invertible sheaf is certainly quasicoherent. Moreover, if $\underline{O}_X$ is coherent, then we may prove that a sheaf of $\underline{O}_X$-modules E on X is invertible if and only if it is coherent and E_x is a free $\underline{O}_{X,x}$-module of rank 1 for all $x \in X$. Indeed, as we have pointed out before, if E is locally finitely presented, then the canonical homomorphism

$$\textbf{Hom}(E, F)_x \rightarrow [E_x, F_x]$$

is bijective for any $F \in \textbf{S}(X, \underline{O}_X)$. From this it follows that if E and F are locally finitely presented and if E_x and F_x are isomorphic for some $x \in X$, then $E|U$ and $F|U$ are isomorphic for some open neighborhood U of x. Indeed, if $u : E_x \rightarrow F_x$ and $v : F_x \rightarrow E_x$ are mutually inverse isomorphisms, then we pick $f \in \textbf{Hom}(E, F)(V)$ resp. $g \in \textbf{Hom}(F, E)(V)$ for some open neighborhood V of x with $f_x = u$ and $g_x = v$, so $(uv)_x$ and $(vu)_x$ reduce to the identity and it follows that for some open neighborhood $U \subset V$ of x, the morphisms $uv|U$ and $vu|U$ are the identity, proving the assertion. Applying this to an isomorphism $E_x \to \underline{O}_{X,x}$ yields the assertion.

More generally, this argument also proves that a coherent sheaf of $\underline{O}_X$-modules E is locally free (of rank n) if and only if E_x is a free $\underline{O}_{X,x}$-module (of rank n) for all x in X.

(7.2.) Proposition Let $(X, \underline{O}_X)$ be a locally ringed space and let E bea sheaf of $\underline{O}_X$-modules, which is locally of finite type, then E is invertible if and only if $E \otimes F$ is isomorphic to $\underline{O}_X$.

Proof Let $x \in X$ and write $R = \underline{O}_{X,x}$, resp. $M = E_x$ and $N = F_x$. If m denotes the unique maximal ideal of R, then from $M \otimes_R N = R$ we deduce that $M/mM \otimes N/mN = R/m$ (tensorproduct over R/m), so $M/mM = R/m$ up to isomorphism. As M is finitely generated, Nakayama's Lemma yields that we may identify M with R. In particular, we may find a section $s \in E(U)$ on some neighborhood U of x such that sending r to $s_x r$ defines an isomorphism of $\underline{O}_{X,x}$ onto E_x. Argueing as usually, we may assume up to reducing U, that there is an epimorphism $\underline{O}_X|U \to E|U$, which yields for every $y \in U$ an isomorphism $\underline{O}_{X,y}/m_y \cong E_y/m_y E_y$, where m_y is the maximal ideal of $\underline{O}_{X,y}$. Essentially applying Nakayama's Lemma again, it follows that the above epimorphism is actually an isomorphism, which finishes the proof.

(7.3.) It follows in particular that the F, whose existence is claimed in the foregoing Proposition, is necessarily (isomorphic to) E^*.

Note also that (7.2.) applies to the case where $(X, \underline{O}_X)$ is a scheme and the case $(Y, O_X|Y)$, where Y is a generically closed subset of X. If $(X, \underline{O}_X)$ is a scheme, then one easily proves that a quasicoherent sheaf of $\underline{O}_X$-modules E on X is invertible if and only if there exists another quasicoherent sheaf of $\underline{O}_X$-modules F qmd an isomorphism $E \otimes F \cong \underline{O}_X$, for in this case one easily reduces to the affine case, where this is just the definition of an invertible module.

(7.4.) Let us assume from now on that X is a separated scheme, which is locally of finite type with respect to some generically closed subset Y. Using the remarks made in the foregoing Section, the reader may wish to replace the separatedness assumption by some finiteness condition. We leave it to him to verify the details involved.

In general, it is usually difficult to study the group $\mathrm{Pic}(Y) = \mathrm{Pic}(Y, \underline{O}_X|Y)$, mainly due to the lack of structure of Y. In order to remedy this, we will relate $\mathrm{Pic}(Y, \underline{O}_Y)$ to a group based upon isomorphism classes of certain quasicoherent sheaves of $\underline{O}_X$-modules on X. The behaviour of these being reasonably well understood in many situations, this will allow us to calculate the group $\mathrm{Pic}(Y, \underline{O}_Y)$ effectively.

(7.5.) A quasicoherent sheaf of $\underline{O}_X$-modules E on X is said to be <u>locally Y-finitely generated</u> or <u>locally of Y-finite type</u> if there exists a covering of X by open affines $U = \mathrm{Spec}(R)$, such that for each of these the R-module $\Gamma(U, E)$ is $\sigma_{U \cap Y}$-finitely generated. If each of these is $\sigma_{U \cap Y}$- finitely presented, then we say that E is <u>locally Y-finitely presented</u> or <u>locally of Y-finite presentation</u>. In particular, if $(X, \underline{O}_X)$ is locally Y-noetherian, then E is locally Y-finitely of finite type if and only if it is locally of Y-finite presentation if and only if it is Y-coherent.

If E is locally of Y-finite type, then the proof of (6.24.) may easily be modified to derive that for any open affine subset $U = \mathrm{Spec}(R)$ of X, the module $\Gamma(U, E)$ is $\sigma_{U \cap Y}$- finitely generated.

It also follows that E|Y is then locally of finite type on Y. Indeed, let $U = \mathrm{Spec}(R)$ be an open affine subset of X, then $E|U = \underline{O}_M$ for some R-module M, which is by assumption $\sigma = \sigma_{U \cap Y}$- finitely generated . Pick a finitely generated R-submodule N of M such that M/N is σ-torsion, then for some positive integer n, there exists a surjective map $u : R^n \to N$. Passing to the associated quasicoherent sheaves and restricting to $U \cap Y$, we find an epimorphism

$$\underline{O}_R{}^n|U \cap Y \to \underline{O}_N|U \cap Y \to 0.$$

But, $\underline{O}_N|U \cap Y = \underline{O}_M|U \cap Y$, since $Q_\sigma(N) = Q_\sigma(M)$ and from this we obtain an epimorphism

$$\underline{O}_Y{}^n|U \cap Y \to E|U \cap Y \to 0,$$

where as usually $\underline{O}_Y = \underline{O}_X|Y$. Since the $U \cap Y$ form a basis for the topology on Y, this proves the assertion.

Of course, one proves similarly that any E which is locally of Y-finite presentation yields a sheaf $E|Y$ on Y, which is locally of finite presentation.

(7.6.) A quasicoherent sheaf of $\underline{O}_X$-modules E on X is said to be Y-<u>inverible</u> if there exists a quasicoherent sheaf of $\underline{O}_X$-modules F together with an isomorphism $Q_Y(E \otimes F) \cong Q_Y(\underline{O}_X)$. If $U = \mathrm{Spec}(R)$, then $E|U = \underline{O}_P$ and $F|U = \underline{O}_Q$ for some R-modules P and Q, so, if $Y \cap U = \mathbf{K}(\sigma)$ for some idempotent kernel functor σ in R-mod, then $Q_\sigma(P \otimes Q) = Q_\sigma(R)$. It follows that $P = \Gamma(U, E)$ is σ-invertible in the sense of [VV2]. This will sometimes permit us to reduce problems to the "local" case, i.e. we may work over a ring and apply the techniques developed in loc.cit.

(7.7.) Proposition If the quasicoherent sheaf of $\underline{O}_X$-modules E on X is Y-invertible, then E is locally of Y-finite type.

Proof Due to the previous remarks, it clearly suffices to prove this in the affine case. Here, the problem reduces to showing that if the idempotent kernel functor σ has finite type in R-mod, then any σ-invertible R-module P is σ-finitely generated.

Let Q be an "inverse" for P, i.e. there is an isomorphism $Q_\sigma(Q \otimes P) = Q_\sigma(R)$, which we view as an identification. Let us denote by $(\,-\,)'$ quotients modulo σ-torsion, then from $1 \in Q_\sigma(R)$, it follows that we may find a finitely generated $I \in \mathbf{L}(\sigma)$ such that $I \subset (Q \otimes P)'$. Put $I = \sum Rx_j$, then each x_j is of the form $x_j = (\sum_i q_{ij} \otimes p_{ij})'$ for some $q_{ij} \in Q$ and $p_{ij} \in P$. Put $P_1 = \sum_{i,j} Rp_{ij}$, then P_1 is a finitely generated submodule of P. Let $f : Q \otimes P_1 \to (Q \otimes P)'$ be the map $Q \otimes P_1 \to Q \otimes P \to (Q \otimes P)'$. Clearly $I \subset \mathrm{Im}(f) \subset (Q \otimes P)'$, and so $\mathrm{Im}(f)/I$ is σ-torsion, as $(Q \otimes P)/I$ is σ-torsion. So $\mathrm{Coker}(Q \otimes P_1 \to Q \otimes P)$ is σ-torsion. By localizing at each prime ideal in $\mathbf{K}(\sigma)$, the σ-invertibility of Q yields that $\mathrm{Ker}(Q \otimes P_1 \to Q \otimes P)$ is σ-torsion, hence $Q_\sigma(Q \otimes P_1) = Q_\sigma(Q \otimes P) = Q_\sigma(R)$. We thus get $Q_\sigma(P_1) = Q_\sigma(R \otimes P_1) = Q_\sigma(Q_\sigma(R) \otimes P_1) = Q_\sigma(Q_\sigma(P \otimes Q) \otimes P_1) = Q_\sigma(P \otimes Q \otimes P_1) = Q_\sigma(P \otimes Q_\sigma(Q \otimes P_1)) = Q_\sigma(P \otimes Q_\sigma(R)) = Q_\sigma(P \otimes R) = Q_\sigma(P)$ and this proves that P/P_1 is σ-torsion, i.e. that P is σ-finitely generated, indeed.

(7.8.) Proposition If X is locally noetherian and E is a quasicoherent sheaf of $\underline{O}_X$-modules, which is locally of Y-finite type, then E is Y-invertible if and only if there exists a sheaf of $\underline{O}_X$-modules F on X such that $\mathrm{Cl}_{X/Z}(E \otimes F)$ is isomorphic to $\mathrm{Cl}_{X/Z}(\underline{O}_X) = Q_Y(\underline{O}_X)$, where $Z = X - Y$, as usually.

Proof We will see in (7.10.) below, that a quasicoherent sheaf of $\underline{O}_X$-modules E on Y is Y-invertible if and only if $E|Y$ is an invertible sheaf of $\underline{O}_X$-modules on Y and from this one may deduce the result by using (7.2.).

Another proof may be given as follows. First, if E is Y-invertible, then by definition there exists a quasicoherent sheaf of $\underline{O}_X$-modules F together with an isomorphism $Q_Y(E \otimes F) \cong Q_Y(\underline{O}_X)$. But from (6.40.), it then follows that $Q_Y(E \otimes F) = \mathrm{Cl}_{X/Z}(E \otimes F)$, whence the assertion. Conversely, since $(H^0_{X/Z}E)_p = E_p$ for all $p \in Y$ by (6.36.), we get that $\mathrm{Cl}_{X/Z}(E \otimes F)_p = E_p \otimes F_p$. Argueing as in (7.2.), it easily may be deduced that for some open neighborhood U of p, which we may choose to be affine, i.e. of the form $U = \mathrm{Spec}(R)$, there exists an epimorphism $\underline{O}_X|U \to E|U$ inducing an isomorphism $\underline{Q}_X | U \cap Y \to E | U \cap Y$. If we let $U \cap Y = \mathbf{K}(\sigma)$ for some idempotent kernel functor σ in R-mod, and if $E|U = \underline{Q}_M$ for some R-module M, then $Q_\sigma(R) = Q_\sigma(M)$, hence M is certainly σ-invertible. We leave it as an exercise to the reader to verify that E is Y-invertible if and only if there exists an open covering by affines U, such that $\Gamma(U, E)$ is $\sigma_{U \cap Y}$-invertible for each of these. From this the result follows easily.

(7.9.) Proposition The isomorphism classes of Y-closed Y-invertible sheaves of $\underline{Q}_X$-modules on X forms a group, which we denote by Pic(X, Y) or Pic(X, $\underline{Q}_X$; Y) and call the <u>relative Picard group</u> of X with respect to Y.

Proof The multiplication in Pic(X, Y) will be given by

$$[E_1] \cdot [E_2] = [Q_Y(E_1 \otimes E_2)],$$

for any pair of Y-closed Y-invertible sheaves of $\underline{Q}_X$-modules E_1, E_2. The only crucial point is the fact that for any of quasicoherent sheaves of $\underline{Q}_X$-modules E, F on X we need

$$Q_Y(E \otimes Q_Y(F)) = Q_Y(E \otimes F) = Q_Y(Q_Y(E) \otimes F).$$

First we reduce to the affine case. If $U = \mathrm{Spec}(R)$ is an affine open subset of X, then $Y \cap U = \mathbf{K}(\sigma)$ for some idempotent kernel functor σ of finite type in R-mod and $E|U = \underline{Q}_M$ resp. $F|U = \underline{Q}_N$ for some R-modules M and N. Denote by Q the localization Q_σ at σ, then it follows that $Q_Y(E)|U = \underline{Q}_{Q(M)}$ and $Q_Y(F)|U = \underline{Q}_{Q(N)}$, so

$$(E \otimes Q_Y(F))|U = \underline{Q}_M \otimes \underline{Q}_{Q(N)} = \underline{Q}_{M \otimes Q(N)}$$

and

$$Q_Y(E \otimes Q_Y(F))|U = \underline{Q}_{Q(M \otimes Q(N))}.$$

Similarly,

$$Q_Y(E) \otimes F)|U = \underline{Q}_{Q(M)} \otimes \underline{Q}_N = \underline{Q}_{Q(M) \otimes N}$$

and

$$Q_Y(Q_Y(E) \otimes F)|U = \underline{Q}_{Q(Q(M) \otimes N)}.$$

The proof of our claim then reduces to the verification of

$$Q_\sigma(Q_\sigma(M) \otimes N) = Q_\sigma(M \otimes N) = Q_\sigma(M \otimes Q_\sigma(N)).$$

But, for example, the canonical map $u : M \otimes N \to Q_\sigma(M) \otimes N$ induced by the localization morphism $j_M : M \to Q_\sigma(M)$ yields for each prime ideal p in $\mathbf{K}(\sigma)$ an isomorphism

$$u_p : (M \otimes N)_p = M_p \otimes N_p \to Q_\sigma(M)_p \otimes N_p = (Q_\sigma(M) \otimes N)_p,$$

proving that u induces an isomorphism $Q_\sigma(M \otimes N) = Q_\sigma(Q_\sigma(M) \otimes N)$. The other isomorphism may be deduced similarly.

(7.10.) Proposition Under the above assumptions, i.e. X is a separated scheme which is locally of finite type with respect to a generically closed subset Y, there is an isomorphism of groups

$$\mathrm{Pic}(X, Y) \cong \mathrm{Pic}(Y, \underline{Q}_X|Y).$$

Proof Let E be a Y-invertible sheaf of $\underline{O}_X$-modules on X, then E is quasicoherent, hence E| Y is quasicoherent on Y. Moreover, E is locally of Y-finite type, hence it follows from (7.5.) that E| Y is locally of finite type. Moreover, by assumption there exists a quasicoherent F on X with $Q_Y(E \otimes F) = Q_Y(\underline{O}_X)$, i.e. such that $E|Y \otimes F|Y = (E \otimes F)|Y = Q_Y(E \otimes F)|Y = Q_Y(\underline{O}_X)|Y = \underline{O}_X|Y = \underline{O}_Y$, so by (7.2.), it follows that E|Y is invertible on Y. Conversely, if E is invertible on Y, then there exists a quasicoherent sheaf of $\underline{O}_Y$-modules f on Y with $E \otimes F = \underline{O}_Y$. Choose E' resp F' in $\mathbf{Q}(X, \underline{O}_X)$ with E'|Y = E resp. F'|Y = F, then we obtain that $Q_Y(E' \otimes F')|Y = E'|Y \otimes F'|Y = E \otimes F = \underline{O}_Y = Q_Y(\underline{O}_X)|Y$, hence $Q_Y(E' \otimes F') = Q_Y(\underline{O}_X)$ and so E' is Y-invertible by definition. Clearly, in the foregoing E' and F' may be chosen to be Y-closed. So, there is a one-to-one correspondence between invertible sheaves of $\underline{O}_Y$-modules on Y and Y-closed Y-invertible sheaves of $\underline{O}_X$-modules on X. From this one easily deduces the isomorphism Pic(X, Y) = Pic(Y). We leave details to the reader.

(7.11.) If [E] denotes the isomorphism class of a Y-closed Y-invertible sheaf on X, then we claim that the element $[E]^{-1} \in$ Pic(X, Y) may be represented by the sheaf **Hom**(E, $Q_Y(\underline{O}_X)$).

Indeed, since E|Y is invertible on Y, we may find open affines U = Spec(R) of X covering Y and such that $E|U \cap Y = \underline{O}_X|U \cap Y$ up to isomorphism. Fix such an open affine U and let $U \cap Y = \mathbf{K}(\sigma)$ for an idempotent kernel functor σ in R-mod, then we have $E|U = \underline{O}_M$ for some σ-closed R-module M and $Q_Y(\underline{O}_X)|U = \underline{O}_S$, where $S = Q_\sigma(R)$, hence $E|U \cap Y = \underline{O}_X|U \cap Y$ implies M = S and so E|U is a free $Q_Y(\underline{O}_X)|U$-module of rank one. But

$$\mathbf{Hom}(E, Q_Y(\underline{O}_X))|U = \mathbf{Hom}(E|U, Q_Y(\underline{O}_X)|U),$$

and this reduces to **Hom**($\underline{O}_M$, $\underline{O}_S$), so it does not make any difference whether we take our maps to be $\underline{O}_X|U$-linear or $Q_Y(\underline{O}_X)|U$-linear. We thus find an isomorphism

$$(E \otimes \mathbf{Hom}(E, Q_Y(\underline{O}_X)))|U = E|U \otimes \mathbf{Hom}(E|U, Q_Y(O_X)|U) = Q_Y(O_X)|U$$

since $\mathbf{Hom}(E|U, E|U) = \mathbf{Hom}(Q_Y(\underline{O}_X)|U, Q_Y(\underline{O}_X)|U) = Q_Y(\underline{O}_X)|U$. Since these U cover Y, this also yields an isomorphism

$$Q_Y(E \otimes \mathbf{Hom}(E, Q_Y(\underline{O}_X)) = Q_Y(\underline{O}_X).$$

To finish the proof, we still have to show that $\mathbf{Hom}(E, Q_Y(\underline{O}_X))$ is Y-closed. Let $U = \mathrm{Spec}(R)$, etc. as before, then

$$\Gamma(U, \mathbf{Hom}(E, Q_Y(\underline{O}_X))|Y) = \mathrm{Hom}(\underline{O}_M, \underline{O}_S) = [M, Q_\sigma(R)].$$

If N is an R-module, then $[N, [M, Q_\sigma(R)]] = [N \otimes M, Q_\sigma(R)]$, so to conclude, it suffices to verify that if $I \in \mathbf{L}(\sigma)$, then $K = \mathrm{Ker}(u : I \otimes M \to R \otimes M)$ is σ-torsion, for then $\mathrm{Ker}(u)$ and $\mathrm{Coker}(u)$ are both torsion and the canonical map $[R \otimes M, Q_\sigma(R)] \to [I \otimes M, Q_\sigma(R)]$ is an isomorphism, hence so is $[E, [M, Q_\sigma(R)] \to [I, [M, Q_\sigma(R)]]$, and $[M, Q_\sigma(R)]$ is σ-closed, indeed. But for each $p \in \mathbf{K}(\sigma)$, we have $I_p = R_p$, hence $K_p = \mathrm{Ker}(u_p : I_p \otimes M_p \to R_p \otimes M_p) = 0$, implying that K is σ-torsion.

(7.12.) Note The constructions and definitions given above may actually be generalized to arbitrary (locally) ringed spaces $(X, \underline{O}_X)$ and generically closed subsets of these by working with the closure operator $\mathrm{Cl}_{X/Z}$, where $Z = X - Y$. A possible approach is to define a quasicoherent sheaf of $\underline{O}_X$-modules E on X to be Y-invertible (or "Z-invertible", depending on your personal taste) if for some F (which has to be assumed quasicoherent if $(X, \underline{O}_X)$ is not locally ringed) we have that

$Cl_{X/Z}(E \otimes F) = Cl_{X/Z}(\underline{O}_X)$. We leave it to the reader to verify that one may use these to construct a group $Pic(X, Y)$ (or $Pic(X, Z)$!) which coincides with the one we defined, e.g. when X is a locally noetherian, separated scheme.

(7.13.) The group $Pic(X, Y)$ behaves functorially with respect to Y. For, if $Y' \supset Y''$, then there is a group morphism $Pic(X, Y') \rightarrow Pic(X, Y'')$.

This map is defined by sending the class of some Y'-closed Y'-invertible E on X to that of $Q_{Y''}(E)$. Clearly $Q_{Y''}(E)$ is Y''-closed and it is Y''-invertible, since

$$Q_{Y''}(E \otimes F) = Q_{Y''}(Q_{Y'}(E \otimes F)) = Q_{Y''}(Q_{Y'}(\underline{O}_X)) = Q_{Y''}(\underline{O}_X),$$

for any F with $Q_{Y'}(E \otimes F) = Q_{Y'}(\underline{O}_X)$. It is easy to see that this map actually coincides (up to isomorphism) with the restriction morphism

$$Pic(Y') \rightarrow Pic(Y'') : [E] \rightarrow [E|Y''],$$

using (7.10.) above. This construction will be generalized below.

(7.14.) Let us call a couple (R, σ), where R is a ring and σ an idempotent kernel functor of finite type in R-mod, a <u>torsion couple</u>. By definition a morphism $(R, \sigma) \rightarrow (S, \tau)$ of torsion couples is a ring morphism $f : R \rightarrow S$ with the property that $f^{-1}(q) \in \mathbf{K}(\sigma)$ for all $q \in \mathbf{K}(\tau)$.

Recall that if $f : R \rightarrow S$ is a ring morphism and if σ is an idempotent kernel functor in R-mod, then the idempotent kernel functor $f_*\sigma$ in S-mod is defined by letting an S-module be $f_*\sigma$-torsion if it is σ-torsion, when considered as an R-module, by restriction of scalars through f.

(7.15.) Lemma Let (R, σ) and (S, τ) be torsion couples, then for any ring morphism $f : R \rightarrow S$ the following assertions are equivalent:

(7.15.1.) f induces a morphism of torsion couples $f : (R, \sigma) \to (S, \tau)$;

(7.15.2.) $f_*\sigma \leq \tau$.

Proof Let $f : (R, \sigma) \to (S, \tau)$ be a morphism of torsion couples, then if q belongs to $\mathbf{K}(\tau)$, it follows that $f^{-1}(q) \in \mathbf{K}(\sigma)$. Now, if $q \in \mathbf{K}(f_*\sigma)$, then of course $\sigma(S/q) \neq 0$, so we may find some $I \in \mathbf{L}(\sigma)$ such that $f(I)s \subset q$ for some $s \notin q$, hence $f(I) \subset q$ i.e. $I \subset f^{-1}(q)$ and we obtain that $f^{-1}(q) \in \mathbf{L}(\sigma)$, a contradiction. Now, σ and τ have finite type, and as we have pointed out before, $f_*\sigma$ has finite type too, hence $f_*\sigma$ and τ are completely determined by $\mathbf{K}(f_*\sigma)$ and $\mathbf{K}(\tau)$, so it follows that $f_*\sigma \leq \tau$, indeed.

Conversely, if $q \in \mathbf{K}(\tau)$, then $\tau(S/q) = 0$, hence $(f_*\sigma)(S/q) = 0$, so we get $\sigma(S/q) = 0$, where S/q is viewed as an R-module. Now, $R/f^{-1}(q)$ injects into S/q, hence $\sigma(R/f^{-1}(q)) = 0$ and $f^{-1}(q) \in \mathbf{K}(\sigma)$.

(7.16.) The global version of this notion works as follows : a couple (X, Y) (or even $(X, \underline{O}_X; Y)$ if ambiguity arises!) is a (global) <u>torsion couple</u> if X (or $(X, \underline{O}_X)$) is a separated scheme and Y a generically closed subset of X with respect to which X is locally of finite type. As usually, the hypothesis of X being separated is not strictly necessary, but we include it for convenience's sake.

Recall that a morphism of schemes $u : X \to X'$ is said to be <u>affine</u> if there exists an open covering of X' by affines V such that $u^{-1}(V)$ is an open affine subset of X for each of these. If $u : X \to X'$ is affine in this sense, then $u^{-1}(V)$ is affine in X for <u>every</u> open affine subset V of X'.

A morphism of torsion couples $(X, Y) \to (X', Y')$ is then by definition just a scheme morphism $u : X \to X'$, which is affine and such that $u(Y) \subset Y'$.

An affine torsion couple is one of the form $(\mathrm{Spec}(R), Y)$ (up to isomorphism in the category of all torsion couples). In this case, Y is necessarily of the form $Y = \mathbf{K}(\sigma)$ for some idempotent kernel functor σ in R-mod, which is of finite type. It follows that the full subcategory consisting of all affine torsion couples may be identified with the dual of the category of torsion couples (R, σ) - we will sometimes call these "local" torsion couples, in contrast to the geometric "global" torsion couples.

(7.17.) Proposition A morphism of torsion couples $u : (X, Y) \to (X', Y')$ induces a group morphism $\mathrm{Pic}(X', Y') \to \mathrm{Pic}(X, Y)$, by sending the element $[E] \in \mathrm{Pic}(X', Y')$ to $[Q_Y(u_*E)] \in \mathrm{Pic}(X, Y)$.

Proof Let E be a Y'-invertible quasicoherent sheaf of $\underline{O}_{X'}$-modules on X' and let $U = \mathrm{Spec}(R)$ be an affine open subset of X', then $u^{-1}(U)$ is an affine open subset of X, say $V = u^{-1}(U) = \mathrm{Spec}(S)$. We then have $E|U = \underline{O}_M$ for some R-module M, which is $\sigma = \sigma_{U \cap Y'}$-invertible and $Q_Y(u^*E)|V = Q_{Y \cap U}(u^*E|V) = Q_{Y \cap V}(u^*(E|U))$. Let $\sigma_{Y \cap V} = \tau$ and let $N = Q_t(M \otimes S)$, then it follows that $Q_Y(u^*E)|V = \underline{O}_N$.

This reduces the problem to showing that if P and Q are σ-invertible R-modules, then

$$Q_\tau(Q_\tau(P \otimes_R S) \otimes_R Q_\tau(Q \otimes_R S)) = Q_\tau(Q_\sigma(P \otimes_R Q) \otimes_R S),$$

for then, if $Q_\sigma(P \otimes Q) \cong Q_\sigma(R)$, then $Q_\tau(Q_\tau(P \otimes S) \otimes Q_\tau(Q \otimes S)) = Q_\tau(Q_\sigma(R) \otimes S) = Q_\tau(S)$ and so $Q_\tau(P \otimes S)$ is τ-invertible, etc.

But, $Q_\tau(Q_\tau(P \otimes S) \otimes Q_\tau(Q \otimes S)) = Q_\tau((P \otimes Q) \otimes S)$ and as $f_*\sigma \leq \tau$, the latter is just

$$Q_\tau(Q_{\sigma'}(P \otimes Q \otimes S)) = Q_\tau(Q_{\sigma'}(Q_\sigma(P \otimes Q) \otimes S) = Q_\tau(Q_\sigma(P \otimes Q) \otimes S)$$

(with $\sigma' = f_*\sigma$). Here, we use the fact that if $f : R \to S$ is a ring morphism and σ an idempotent kernel functor in R-mod, then for any S-module M, the R-module $Q_\sigma(_RM)$ is canonically endowed with an S-module structure and for this S-module structure the modules $Q_{\sigma'}(M)$ and $Q_\sigma(_RM)$ are canonically isomorphic. In particular, the rings $Q_\sigma(_RS)$ and $Q_{\sigma'}(S)$ are isomorphic.

(7.18.) Corollary If $f : (R, \sigma) \to (S, \tau)$ is a morphism of torsion couples, then f induces a group morphism $Pic(R, \sigma) \to Pic(S, \tau)$.

The foregoing results generalize (7.13.).

(7.19.) Example Let $f : R \to S$ be a ring morphism and let I and J be ideals of R and S. Denote by $X_R(I)$ resp. $X_S(J)$ the corresponding open subsets of Spec(R) resp. Spec(S). Assume that I and J are finitely generated then we get associated idempotent kernel functors σ_I resp. σ_J in R-mod resp. S-mod. The morphism f yields a morphism of torsion couples $f : (R, \sigma_I) \to (S, \sigma_J)$ if and only if $f_*\sigma_I \leq \sigma_J$. Now $f_*\sigma_I = \sigma_{(f(I))}$, where $(f(I))$ is the S-ideal generated by f(I), and $\sigma_{(f(I))} \leq \sigma_J$ is equivalent to $X_S(J) \subset X_S(f(I))$, hence also to f inducing a morphism $Spec(S) \to Spec(R)$, which maps $X_S(J)$ into $X_R(I)$. In this case, the above results say that there is an induced map $Pic(R, \sigma_I) \to Pic(S, \sigma_J)$.
As we know that $Pic(R, \sigma_I) = Pic(X_R(I))$ and $Pic(S, \sigma_J) = Pic(X_S(J))$, this map reduces to the morphism $Pic(X_R(I)) \to Pic(X_S(J))$ given by sending the class [E] of an invertible sheaf E on $X_R(I)$ to its "restriction" $[(^af)^*(E)]$ in $Pic(X_S(J))$.

(7.20.) Note Let $(X, Y) \to (X', Y')$ be a morphism of torsion couples, then the induced map $Pic(X', Y') \to Pic(X, Y)$ may also be given a K-theoretic interpretation.
Indeed, the category **Pic**(X, Y) of Y-closed Y-invertible sheaves of $\underline{O}_X$-modules is a category with product, this product being given by $E \perp F = Q_Y(E \otimes F)$ for any $E, F \in$ **Pic**(X, Y). The K-groups K_0**Pic**(X, Y) and K_1**Pic**(X, Y) are defined in [Ba] for example, and one easily verifies that K_0**Pic**$(X, Y) = Pic(X, Y)$.

On the other hand, in the terminology of loc.cit., the full subcategory $<Q_Y(\underline{O}_X)>$ of $\underline{\textbf{Pic}}(X, Y)$ consisting of the only object $Q_Y(\underline{O}_X)$ is cofinal in $\underline{\textbf{Pic}}(X, Y)$. Hence, the inclusion functor $<Q_Y(\underline{O}_X)> \rightarrow \underline{\textbf{Pic}}(X, Y)$ induces an isomorphism $K_1\underline{\textbf{Pic}}(X, Y) \cong K_1<Q_Y(\underline{O}_X)>$ and it is easy to see that $K_1<Q_Y(\underline{O}_X)> = \mathrm{Aut}(Q_Y(\underline{O}_X)) = \Gamma(Y, (\underline{O}_X|Y)^*)$. Let us denote this group by $U(X, Y)$.

If $u : (X, Y) \rightarrow (X', Y')$ is a morphism of torsion couples, then it induces a functor

$$\underline{\textbf{Pic}}(X', Y') \rightarrow \underline{\textbf{Pic}}(X, Y)$$

which is product preserving. It then follows from the (K_0, K_1)-exact sequence, that we have an exact sequence

$$U(X', Y') \rightarrow U(X, Y) \rightarrow \mathrm{Pic}(u) \rightarrow \mathrm{Pic}(X', Y') \rightarrow \mathrm{Pic}(X, Y),$$

cf. [Ba]. The intermediate group $\mathrm{Pic}(u)$ is defined K-theoretically and is studied in the present context (in the affine case, however!), in [VV2].

(7.21.) Let $(X, \underline{O}_X)$ be an arbitrary ringed space, then as in [Au1] we define a sheaf of $\underline{O}_X$-modules E on X to be locally projective of finite type if it is locally a direct summand of a (locally) free sheaf of $\underline{O}_X$-modules of finite rank, i.e. for each $x \in X$ there should exist an open neighborhood U of x and sheaf morphisms of the form $u : E|U \rightarrow (\underline{O}_X)^n|U$ resp. $v : (\underline{O}_X)^n|U \rightarrow E|U$ such that vu is the identity on $E|U$. It has been proved in loc.cit. that this is equivalent to E being locally presented and E_x being a projective $\underline{O}_{X,x}$-module for all $x \in X$. If $\underline{O}_X$ is a sheaf of local rings, then an argument as in (7.1.) shows that E is then locally free and conversely.

(7.22.) Under the same assumptions, for any sheaf of $\underline{O}_X$-modules E, there is a natural morphism of sheaves of rings $\omega : \underline{O}_X \rightarrow \textbf{End}(E)$, which "locally" maps r in $\underline{O}_X$ to the multiplication by r in E. If $\mathrm{Ker}(\omega) = 0$, then we call E <u>faithful</u>. In particular, if E is locally finitely presented, then it is faithful if and only if E_x is a faithful $\underline{O}_{X,x}$-module for every $x \in X$.

We now call a sheaf of $\underline{O}_X$-modules E a <u>progenerator</u> if it is faithful and locally projective of finite type.

(7.23.) Let (X, Y) be a torsion couple, then we say that a sheaf of $\underline{O}_X$-modules E is Y-<u>faithful</u> if E|Y is a faithful sheaf of $\underline{O}_X|Y = \underline{O}_Y$-modules. If E is locally Y-finitely presented, then this means that E_x is faithful for all x in X.

The corresponding local notion corresponds to a torsion couple (R, σ), where we say that an R-module M is σ-<u>faithful</u> if and only if the sheaf $\underline{O}_M$ is $\mathbf{K}(\sigma)$-faithful in the previous sense. Clearly, if M is σ-finitely presented, then M is σ-faithful if and only if M_p is a faithful R_p-module for all $p \in \mathbf{K}(\sigma)$. Since $\mathrm{Ker}(R \rightarrow [M, M]) = \mathrm{Ann}_R(M)$, it trivially follows that M is then σ-faithful if and only if $\mathrm{Ann}_R(M)$ is σ-torsion.

We call a sheaf of $\underline{O}_X$-modules E on X a Y-<u>progenerator</u> if it is Y-closed, locally Y-finitely presented and if for all $x \in X$ the module E_x is free (of finite rank) over $\underline{O}_{X,x}$. This terminology is somewhat misleading, since in general there is no natural category in which E is a progenerator in the category theoretic sense.

It is clear that any Y-progenerator is also Y-faithful, hence it is fairly easy, using the same method as in (7.10.), to prove the following result:

(7.24.) Proposition Let X be a separated scheme, which is locally of finite type with respect to a generically closed subset Y, then there is a one-to-one correspondence between:

(7.24.1.) Y-progenerators on $(X, \underline{O}_X)$;

(7.24.2.) locally projective sheaves of finite type on $(Y, \underline{O}_Y)$;

(7.24.3.) locally free sheaves of finite rank on $(Y, \underline{O}_Y)$.

(7.25.) Corollary Under these assumptions, any Y-closed Y-invertible sheaf of $\underline{O}_X$-modules on X is a Y-progenerator.

(7.26.) Let (R, σ) be a torsion couple, then in the terminology of [VV2], we call an R-module M σ-<u>quasiprojective</u> if M_p is a free R_p-module for all p ∈ **K**(σ). As in loc.cit. we then call an R-module M a σ-<u>progenerator</u> if it is σ-closed, σ-finitely presented and σ-quasiprojective. A σ-progenerator is automatically σ-faithful. Since the notion of being a Y-progenerator for some torsion couple (X, Y) is clearly local on X, it follows that a quasicoherent sheaf of $\underline{O}_X$-modules E on X is a Y-progenerator for any affine open subset U of X.

We have the following result, which applies in particular to σ-progenerators over R:

(7.27.) Proposition A σ-finitely presented R-module M is σ-quasiprojective if and only if [M , -] is an exact endofunctor in (R, σ)-mod.
Proof Using (5.28.) and the same argument as in the last part of the proof of (7.11.) one easily shows that [M, N] is σ-closed, whenever the module M is σ-quasiprojective and N is σ-closed, and that [M , -] is then exact in (R, σ)-mod, using the exactness of [M_p, -] in R_p-mod.
Conversely, assume that [M, -] is an exact endofunctor of (R, σ)-mod, and consider an exact sequence

$$0 \to N' \to N \to N'' \to 0$$

in R_p-mod, which is clearly also exact in (R, σ)-mod. So,

$$0 \to [M, N'] \to [M, N] \to [M, N''] \to 0$$

is then also exact in (R, σ)-mod. Applying (5.28.) again, we derive an exact sequence of R_p-modules

$$0 \to [M_p, N'] \to [M_p, N] \to [M_p, N''] \to 0,$$

i.e. $[M_p, -]$ is exact in R_p-mod and so M_p is projective (hence free!).

We leave it as an exercise to the reader to derive a global version of this. Note already that it follows from the proof of (7.26.) that the R-module $End_R(M)$ is σ-closed for any σ-progenerator M in R-mod, hence

(7.28.) Corollary If E is a Y-progenerator on X, then the sheaf **End**(E) is Y-closed.

(7.29.) If E and F are quasicoherent sheaves of $\underline{O}_X$-modules, then we denote by $E \perp F$ the sheaf $Q_Y(E \otimes F)$, at least when no ambiguity arises. Generalizing the terminology of [LO1] or [VV2], we call this the <u>modified tensorproduct</u> of E and F (with respect to Y).

The operation $\perp$ possesses many nice features, which are similar to those of the usual tensorproduct and which we leave to the reader to verify. Here we will mainly be concerned with the behaviour of Y-progenerators with respect to the operations $\perp$ and **Hom**.

Recall that if M and N are R-progenerators, then so is $M \otimes N$ and we have that $End_R(M) \otimes End_R(N) = End_R(M \otimes N)$. Moreover, for any progenerator E we have $M^* \otimes N = End_R(M)$. These results may be generalized as follows:

(7.30.) Proposition Let (X, Y) be a torsion couple and assume that E, F are Y-progenerators, then

(7.30.1.) $E \perp F$ is a Y-progenerator;

(7.30.2.) **End**(E) $\perp$ **End**(F) = **End**(E $\perp$ F);

(7.30.3.) $E^* \perp E = $ **End**(E).

Proof Since for any open subset U of X, we have **Hom**(E, F)|U = **Hom**(E|U, F|U) resp. (E $\perp$ F)|U = (E|U) $\perp$ (F|U), we may assume X to be affine, i.e. of the form Spec(R) for some ring R and Y = **K**(σ) for some idempotent kernel functor σ of finite type. We then have that E = $\underline{Q}_M$ resp. F = $\underline{Q}_N$ for some σ-progenerators M and N. By using the property that M is σ-finitely presented and N is σ-closed, we get that **Hom**($\underline{Q}_M$, $\underline{Q}_N$) = $\underline{Q}_{[M, N]}$, so in particular **End**($\underline{Q}_M$) = $\underline{Q}_{End(M)}$ and **End**($\underline{Q}_N$) = $\underline{Q}_{End(N)}$.

It is easy to check that $Q_\sigma(M \otimes N)$ = M $\perp$ N is a σ-progenerator and that E $\perp$ F = $\underline{Q}_{M \perp N}$. In particular, **End**(E $\perp$ F) = $\underline{Q}_{End(M \perp N)}$. The result now follows from (III.1.6.) and (III.1.9.) in [VV2].

Let us recall the proof of (2) to give an idea of the methods involved.

We have to show that the obvious map

$$u : End_R(M) \otimes End_R(N) \to End_R(M \otimes N) \to End_R(M \perp N)$$

yields an isomorphism $End_R(M) \perp End_R(N)$ = $End_R(M \perp N)$. Let p $\in$ **K**(σ), then $End(M)_p$ = $End(M_p)$ resp. $End(N)_p$ = $End(N_p)$. On the other hand,

$$End(M \perp N)_p = End((M \perp N)_p) = End(M_p \otimes N_p),$$

so u yields, by localizing, an isomorphism at each p $\in$ **K**(σ). Hence $Q_\sigma(u)$ is an isomorphism of the desired type.

The proof of (3) is similar.

(7.31.) Let us call a torsion couple (R, σ) <u>noetherian</u> if R is σ-noetherian; in the global version we call a torsion couple (X, Y) <u>noetherian</u> if X is locally Y-noetherian. Clearly, (X, Y) is noetherian if and only if we may find an open affine covering of X with affines U = Spec(R), such that (R, $\sigma_{U \cap Y}$) is noetherian for each U. In this case for each open affine of the form U = Spec(R), we have that (R, $\sigma_{U \cap Y}$) is noetherian.

If (X, Y) is a noetherian torsion couple, then for any pair of Y-progenerators E and F, we have that **Hom**(E, F) is a Y-progenerator as well. Indeed, let us reduce to the affine situation, so we have to prove that if M and N are σ-progenerators for some σ in R-mod, such that R is σ-noetherian, then the R-module $[E, F]$ is a σ-progenerator as well.

First, since E is σ-finitely generated (or σ-finitely presented, since we assume that R is σ-noetherian) there exists a finitely generated R-submodule E' of E such that E/E' is σ-torsion. It follows that $[E, F] = [E', F]$, since F is σ-closed by assumption. Moreover, there exists some surjective map $u : R^n \to E'$, for some positive integer n and this map induces an inclusion

$$[E', F] \subset [R^n, F] = F^n.$$

Since F is σ-finitely presented, it is σ-noetherian, hence $[E', F]$ is σ-finitely generated, by (5.27.1.), hence $[E, F]$ is σ-finitely presented. By (7.26.), it follows that $[E, F]$ is σ-closed. Finally, for any $p \in \mathbf{K}(\sigma)$, we have that $[E, F]_p = [E_p, F_p]$, hence $[E, F]$ is σ-quasiprojective since the modules E_p and F_p are R_p-progenerators for all $p \in \mathbf{K}(\sigma)$. This proves that $[E, F]$ is a σ-progenerator, indeed.

(7.32.) For any ringed space $(Z, \underline{O}_Z)$, let $\wp(Z, \underline{O}_Z)$ or $\wp(Z)$ denote the full subcategory of $\underline{S}(Z, \underline{O}_Z)$ of all locally free sheaves of $\underline{O}_Z$-modules on Z. Let $\wp(X, Y)$ be the category of Y-progenerators on X, then (7.23.) says that $\wp(X, Y)$ and $\wp(Y)$ may be identified.

Let $Y' \subset Y$, then localizing yields a functor $\wp(X, Y) \to \wp(X, Y')$, which may be identified (up to isomorphism) with the restriction $\wp(Y) \to \wp(Y')$. If (X, Y) is a (global) noetherian torsion couple, then this functor restricts to the functor **Pic**$(X, Y) \to$ **Pic**(X, Y'), defined in (7.20.).

More generally, any morphism of torsion couples $u : (X, Y) \to (X', Y')$ induces a functor $\wp(X', Y') \to \wp(X, Y)$. To prove this, denote by $\wp(R, \sigma)$ etc. the category of σ-progenerators in R-mod. We then have:

(7.33.) Lemma Any morphism of torsion couples $f : (R, \sigma) \to (S, \tau)$ induces a functor

$$\wp(R, \sigma) \to \wp(S, \tau) : P \to Q_\tau(P \otimes_R S).$$

Proof Since P is σ-finitely presented, there exists a finitely presented Q and an R-linear map $u : Q \to P$ with σ-torsion kernel and cokernel. Consider the map $u \otimes S : Q \otimes S \to P \otimes S$ and let $q \in \mathbf{K}(\tau)$, then $f^{-1}(q) = p \in \mathbf{K}(\sigma)$ and so $(Q \otimes S)_q = Q_p \otimes R_p \otimes S_q = P_p \otimes R_p \otimes S_q = (P \otimes S)_q$, from which it follows that $u \otimes S$ has τ-torsion kernel and cokernel, so $P \otimes S$ and $Q_\tau(P \otimes S)$ are τ-finitely presented, since $Q \otimes S$ is clearly finitely presented over S. On the other hand, for the same q and p, the module P_p is free over R_p, since P is a σ-progenerator, hence $P_p \otimes R_p$ is free over S_p, and so is $(P \otimes S)_q = P_p \otimes S_p \otimes S_q$ over S_q. As this holds for all q in $\mathbf{K}(\sigma)$ and as $Q_\tau(P \otimes S)_q = (P \otimes S)_q$, this shows that $Q_\tau(P \otimes S) \in \wp(S, \tau)$.

(7.34.) Proposition Any morphism of torsion couples $f : (R, \sigma) \to (S, \tau)$, where S is τ-closed, induces a functor

$$\wp(R, \sigma) \to \wp(S, \tau) : P \to Q_{\sigma'}(P \otimes_R S) = Q_\sigma(P \otimes_R S) = P \perp S,$$

where $\sigma' = f_* \sigma$.

Proof As (7.33.) has already been proved, it clearly suffices to verify that $Q_\sigma(P \otimes S) = Q_\tau(P \otimes S)$, for any σ-progenerator P. Now, $Q_\tau(S \otimes P) = Q_\tau(Q_{\sigma'}(S \otimes P)) = Q_{\sigma'}(Q_\tau(S \otimes P))$, so it is sufficient to check that the localization map

$$S \otimes_R P \to Q_\tau(S \otimes_R P)$$

has σ-torsion kernel and cokernel. But, for any $p \in \mathbf{K}(\sigma)$, we have that $(S \otimes P)_p = S_p \otimes P_p \cong S_p \otimes R_p{}^n = S_p{}^n$, for some positive integer n, since P_p is a free R_p-module of finite rank, by assumption. Similarly, $Q_\tau(S \otimes P)_p = Q_\tau(S_p \otimes P_p) \cong Q_\tau(S_p \otimes R_p{}^n) = Q_\tau(S_p{}^n) = (Q_\tau(S)_p)^n = S_p{}^n$ and from this one easily deduces that the induced map

$$(S \otimes P)_p \to Q_\tau(S \otimes P)_p$$

is an isomorphism for any $p \in \mathbf{K}(\sigma)$, whence the assertion.

This implies if $\sigma \leq \tau$ and R is τ-closed, then $\wp(R, \sigma) \subset \wp(R, \tau)$.

The same method of proof yields the following, somewhat stronger result:

(7.35.) Corollary If $f : (R, \sigma) \to (S, \tau)$ is a morphism of torsion couples such that S is τ-closed, and if P is an $f_*\sigma$-finitely generated $f_*\sigma$-closed S-module, such that P_p is a projective S_p-module for all $p \in \mathbf{K}(\sigma)$, then P belongs to $\wp(S, \tau)$.

(7.36.) Corollary Any morphism of torsion couples $u : (X, Y) \to (X', Y')$ induces a functor

$$\wp(X', Y') \to \wp(X, Y) : E \to Q_Y(u^*E).$$

If $\underline{Q}_X$ is Y-closed, then for each $E \in \wp(X', Y')$, we have $Q_Y(u^*E) = Q_Z(u^*E)$, where $Z = u^{-1}(Y') \subset X$.

The next Lemmas will be used in (7.53.) below:

(7.37.) Lemma Let $\sigma \leq \tau$ be idempotent kernel functors of finite type in R-mod, where R is assumed to be τ-closed and σ-noetherian. If M is a τ-progenerator, then M is σ-finitely generated.

Proof Since M is τ-finitely generated, we may find a finitely generated R-submodule N of M such that M/N is τ-torsion. Since R is τ-closed, we have $N^* = M^*$, where $(\,\text{-}\,)^* = [\,\text{-}\,, R]$. Choose a positive integer n such that there is a surjective map $R^n \to N$, then dualizing yields an inclusion $M^* = N^* \subset R^n$, so M^* and similarly M^{**} are σ-finitely generated, since R is σ-noetherian by assumption. Since M is τ-finitely presented, we have $(M^{**})_p = (M_p)^{**} = [[M_p, R_p], R_p]$, for all $p \in \mathbf{K}(\tau)$ and since the R-module M is τ-quasiprojective, it follows that the canonical map $M_p \to (M^{**})_p$ is an isomorphism for all of these p. Finally, since both M and M^{**} are τ-closed (by (7.27.)) we obtain that $M = M^{**}$, proving the assertion. (See also a similar statement in the noetherian case in [VV2]!).

(7.38.) Corollary Let M and P be σ-progenerators and let N be a τ-progenerator such that $Q_\tau(M \otimes N) = Q_\tau(P)$, then (still under the same assumptions as above) N is a σ-progenerator.

Proof The foregoing Lemma implies that N is σ-closed and σ-finitely generated and hence σ-finitely presented. We have to verify that N_p is a free R_p-module for all $p \in \mathbf{K}(\sigma)$. Since M and P are σ-progenerators, for each $p \in \mathbf{K}(\sigma)$ we may find positive integers m and n together with isomorphisms $M_p = R_p{}^m$ resp. $P_p = R_p{}^n$. But then

$$Q_\tau(P)_p = Q_\tau(P_p) = Q_\tau(R_p{}^n) = (Q_\tau(R)_p)^n = R_p{}^n$$

and

$$Q_\tau(M \otimes N)_p = Q_\tau(M_p \otimes N_p) = Q_\tau(R_p{}^m \otimes N_p) = Q_\tau(N_p{}^m) = (Q_\tau(N)_p)^m = N_p{}^m.$$

We thus find $N_p{}^m = R_p{}^n$, so N_p is projective, hence free over R_p. This finishes the proof.

(7.39.) Corollary Assume (X, Y) is a noetherian torsion couple such that $\underline{O}_X$ is Y'-closed for some generically closed subset $Y' \subset Y$. If F is a Y'-progenerator and if E and G are Y-progenerators on X with the property that that $Q_{Y'}(E \otimes F) = Q_{Y'}(G)$, then F is a Y-progenerator.

(7.40.) Corollary Under the same assumptions, for any Y-progenerator E the Y'-progenerator $Q_{Y'}(E)$ is a Y-progenerator too.

(7.41.) Let us call a sheaf of $\underline{O}_X$-algebras A on X a Y-<u>Azumaya algebra</u> if it is a Y-progenerator and if the canonical map

$$A \perp A^{\text{opp}} \to \mathbf{End}(A)$$

deduced from $m : A \otimes A^{\text{opp}} \to \mathbf{End}(A)$ is an isomorphism. Here, the "opposite" sheaf of $\underline{O}_X$-algebras A^{opp} of A is defined by putting $\Gamma(U, A^{\text{opp}}) = \Gamma(U, A)^{\text{opp}}$ for every open subset U of X and $m : A \otimes A^{\text{opp}} \to \mathbf{End}(A)$ is locally defined by $m(a \otimes b)(x) = axb$, for any $a, b, x \in A(U)$ and by passing to the associated sheaves afterwards. Of course, (7.28.) implies $\mathbf{End}(A)$ to be Y-closed, since A is assumed to be a Y-progenerator on X. We call $A \otimes A^{\text{opp}}$ resp. $A \perp A^{\text{opp}}$ the <u>enveloping</u> resp. Y-<u>enveloping algebra</u> of A.

Clearly a Y-closed locally Y-finitely presented sheaf of $\underline{O}_X$-algebras A on X is a Y-Azumaya algebra if and only if A_x is an $\underline{O}_{X,x}$-Azumaya algebra (in the usual sense, cf. [DI] or [KO1]) for any $x \in Y$ or equivalently, if A|Y is an Azumaya Algebra on $(Y, \underline{O}_X|Y)$ in the sense of [Au1] or [Gr2].

It is also trivially checked that the notion of being a Y-Azumaya algebra is local, in the sense that a sheaf of $\underline{O}_X$-algebras A is a Y-Azumaya algebra if and only if for all open subsets U of X the sheaf of $\underline{O}_X|U$-algebras A|U is a $Y \cap U$ - Azumaya algebra on U. This is also equivalent to this property holding for any U in an open covering or a basis for X.

Recall from [VV2] that an R-algebra A is said to be a σ-<u>Azumaya algebra</u> with respect to some idempotent kernel functor σ in R-mod if it is a σ-progenerator and if the canonical map $A \otimes A^{\mathrm{opp}} \to \mathbf{End}_R(A)$ yields an isomorphism $A \perp A^{\mathrm{opp}} \cong \mathbf{End}_R(A)$. Again, assuming σ to have finite type, A to be σ-closed and σ-finitely presented as an R-module, this is equivalent to A_p being an Azumaya algebra over R_p, for all $p \in \mathbf{K}(\sigma)$. It thus follows easily that if (X, Y) is a torsion couple and A a quasicoherent sheaf of $\underline{O}_X$-algebras, then A is a Y-Azumaya algebra on X if and only if $\Gamma(U, A)$ is a $\sigma_{U \cap Y}$-Azumaya algebra over R for all open affines U = Spec(R) in X. This is also equivalent to this property being valid for all subsets U in an open affine covering of X. This permits us to prove several results by local-global methods.

(7.42.) Proposition If (X, Y) is a noetherian torsion couple, then **End**(E) is a Y-Azumaya algebra for any Y-progenerator E on X.

Proof It follows from (7.31.) that **End**(E) is a Y-progenerator. To prove the result, reduce to the affine case, i.e. let M be a σ-progenerator over R, where the ring R is σ-noetherian with respect to some idempotent kernel functor σ, then we want to show that $\mathrm{End}_R(M)_p$ is an Azumaya algebra over R_p for all $p \in \mathbf{K}(\sigma)$. But, $\mathrm{End}_R(M)_p = \mathrm{End}(M_p)$, since M is σ-finitely presented and σ-closed, and the latter is an Azumaya algebra over R_p, since M_p is a free R_p-module of finite rank, cf. [AG2]. This finishes the proof.

(7.43.) It is clear that if A and B are Y-Azumaya algebras on X, then so is
A $\perp$ B. Let us say that A and B are Y-<u>similar</u> if there exist Y-progenerators
E and F together with an isomorphism

$$A \perp \mathbf{End}(E) \cong B \perp \mathbf{End}(F).$$

It follows from (7.30.) that this defines an equivalence relation in the set of
all (isomorphism classes of) Y-Azumaya algebras, which we denote by ~
The set of equivalence classes [A] of Y-Azumaya algebras A on X for this
relation is denoted by Br(X, Y) or Br(X, $\underline{O}_X$; Y), when ambiguity may arise.
We call this the <u>relative Brauer group</u> of X with respect to Y.
The multiplication in Br(X, Y) is defined by [A] . [B] = [A $\perp$ B] for any pair of
Y-Azumaya algebras A, B. This is well-defined, since (7.30.) implies that
A ~ A' and B ~ B' yields that A $\perp$ B ~ A' $\perp$ B'. Note also that [$Q_Y(\underline{O}_X)$] plays
the role of identity element for this multiplication and that $Q_Y(\underline{O}_X)$ ~
End(E) for any Y-progenerator E, when X is locally Y-noetherian.
If A is a Y-Azumaya algebra, then A $\perp$ A^{opp} $\cong$ **End**(A) implies that [A]$^{-1}$ =
[A^{opp}] for any Y-Azumaya algebra A on X.
We also have:

(7.44.) Proposition If X is a separated scheme which is locally of finite
type with respect to a generically closed subset of Y, then there is an
isomorphism of groups

$$Br(X, Y) \cong Br(Y, \underline{O}_X|Y).$$

The proof of this result is as in (7.10.); we leave details to the reader.
Here Br(Y, $\underline{O}_X|Y$) is the Brauer group of the ringed space (Y, $\underline{O}_X|Y$),
defined in [Au1], [Gr2], ...

(7.45.) **Note** If (X, Y) is an affine torsion couple, i.e. of the form $(\mathrm{Spec}(R), \mathbf{K}(\sigma))$ for some idempotent kernel functor σ in R-mod (of finite type, as always), then this reduces to the relative Brauer group introduced and studied in [VV2]. Indeed, in this case we have

$$\mathrm{Br}(\mathrm{Spec}(R), \mathbf{K}(\sigma)) = \mathrm{Br}(R, \sigma),$$

with notations as in loc. cit.

(7.46.) **Proposition** Any morphism between noetherian torsion couples $u : (X, Y) \to (X', Y')$ induces a group map $\mathrm{Br}(X', Y') \to \mathrm{Br}(X, Y)$.

Proof We have seen in (7.36.) that u yields a functor $\wp(X', Y') \to \wp(X, Y)$ which sends a Y'-progenerator E to the Y-progenerator $Q_Y(u^*E)$. Restricting to affines, this functor is induced by the one which sends M in $\wp(R, \sigma)$ to $Q_\tau(M \otimes S)$ in $\wp(S, \tau)$ for any morphism $f : (R, \sigma) \to (S, \tau)$ of noetherian torsion couples. This functor sends a σ-Azumaya algebra A to an S-algebra $Q_\tau(A \otimes S)$, which is an S-progenerator. Moreover, if q belongs to $\mathbf{K}(\tau)$, then $f^{-1}(q) = p \in \mathbf{K}(\sigma)$ and it follows that

$$Q_\tau(A \otimes S)_q = (A \otimes S)_q = (A_p \otimes S_p) \otimes S_q$$

is an S_q-Azumaya algebra, since A_p is an R_p-Azumaya algebra. It follows that $Q_\tau(A \otimes S)$ is a τ-Azumaya algebra over S.

Hence, for any Y'-Azumaya algebra A on X', we have proved that $Q_Y(u^*A)$ is a Y-Azumaya algebra on X. We leave it as an easy exercise to the reader that this functor respects the similarity relation $\sim$, from which we derive easily that it induces a map $\mathrm{Br}(X', Y') \to \mathrm{Br}(X, Y)$. Finally, it follows from $u^*(E \otimes F) = u^*E \otimes u^*F$ for any pair $E, F \in \wp(X', Y')$, that $Q_Y(Q_Y(u^*E) \otimes Q_Y(u^*F)) = Q_Y(u^*(E \otimes F)) = Q_Y(u^*(Q_{Y'}(E \otimes F)))$, hence that $\mathrm{Br}(X', Y') \to \mathrm{Br}(X, Y)$ is a group morphism, indeed.

(7.47.) Mimicking some ideas due to [Au1, LO1, Or, Yu] to which we will come back in the next Section, we will now introduce for any pair of generically closed subsets $Y' \subset Y$ of a separated scheme X, which is locally noetherian with respect to Y (and hence with respect to Y') and such that $\underline{O}_X$ is Y'-closed, an intermediate relative invariant $Bcl(X; Y, Y')$.

Let Γ be the set of (isomorphism classes of) objects E in $\wp(X, Y')$ such that $\mathbf{End}(E)$ is a Y-Azumaya algebra on X. This makes sense, as $\mathbf{End}(E)$ is Y'-closed, hence Y-closed. We define an equivalence relation $\sim$ on Γ by putting $E \sim F$ if and only if we may find P and Q in $\wp(X, Y)$, such that $Q_{Y'}(E \otimes P) \cong Q_{Y'}(F \otimes Q)$. It is easy to check, as for the relative Brauer group, that the modified tensor product $\perp$ with respect to Y' defines an abelian group structure on $\Gamma/\sim$. The group thus obtained is denoted by $Bcl(X; Y, Y')$ and called the <u>relative Brauer class group</u> of X (with respect to Y and Y'). The inverse of a class $[E] \in Bcl(X; Y, Y')$ is given by $[E]^{-1} = [E^*]$, where $E^* = \mathbf{Hom}(E, \underline{O}_X)$. Note that we used the noetherian hypothesis here, to infer that E^* is also in $\wp(X, Y')$.

There are maps $i : Pic(X, Y) \to Pic(X, Y')$ and $\beta : Br(X, Y) \to Br(X, Y')$, deduced from (7.13.) and (7.46.). On the other hand, we may also define morphisms $j : Pic(X, Y') \to Bcl(X; Y, Y')$ and $\alpha : Bcl(X; Y, Y') \to Br(X, Y)$. The morphism j sends the element $[M] \in Pic(X; Y')$ to $[M] \in Bcl(X; Y, Y')$ and α sends $[E] \in Bcl(X; Y, Y')$ to $[\mathbf{End}(E)] \in Br(X, Y)$.

We want to prove that the following sequence is exact

$$Pic(X, Y) \to Pic(X, Y') \to Bcl(X; Y, Y') \to Br(X, Y) \to Br(X, Y'),$$

where the maps are the previous ones. We need some preparatory results first. The following results generalize similar statements in [VV2].

(7.48.) Let (R, σ) be a torsion couple and S an arbitrary (not necessarily commutative) R-algebra. Let M be an R-S-bimodule, i.e. M is a left R-

module and a right S-module such that $(rm)s = r(ms)$ for any $r \in R$, $s \in S$ and $m \in M$, then for any left S-module N, we write $M \perp_S N$ for the R-module $Q_\sigma(M \otimes_S N)$. If M is an S-bimodule, then $M \perp_S N$ is necessarily an S-module too.

Assume E is an σ-progenerator and $E^* = [E, Q_\sigma(R)]$, then we write $B = \operatorname{End}_R(E) = E \perp E^*$. We claim that there is an isomorphism $E^* \perp_B E \cong Q_\sigma(R)$, induced by $g : E^* \otimes_B E \to Q_\sigma(R) : \alpha \otimes e \to \alpha(e)$. Indeed, for all p in $\mathbf{K}(\sigma)$ we have $(E^*)_p = E_p{}^* = [E_p, R_p]$ and $B_p = \operatorname{End}_R(E)_p = \operatorname{End}_R(E_p)$, hence $g_p : E_p{}^* \otimes E_p = (E^* \otimes_B E)_p \to Q_\sigma(R)_p = R_p$ is an isomorphism, by a related result in [AG1], so, localizing g at σ we find $E^* \perp_B E \cong Q_\sigma(R)$, indeed.

Let us write (B, R, σ)-mod for the category of left B-modules which are σ-closed as R-modules, viewed as a full subcategory of B-mod. We may view (B, R, σ)-mod as a subcategory of (B, σ')-mod, where σ' is the idempotent kernel functor in B-mod induced by σ.

(7.49.) Proposition Let E be a σ-progenerator for R and $B = \operatorname{End}_R(E)$, then the functor $E \perp - : (R, \sigma)\text{-mod} \to (B, R, \sigma)\text{-mod}$ is an equivalence of categories, with inverse $E^* \perp_B - : (B, R, \sigma)\text{-mod} \to (R, \sigma)\text{-mod}$.

Proof If $L \in (R, \sigma)\text{-mod}$, then $E^* \perp_B (E \perp L) = (E^* \perp_B E) \perp L = Q_\sigma(R) \perp L = L$. Conversely, if $M \in (B, R, \sigma)\text{-mod}$, then $E \perp (E^* \perp_B M) = (E \perp E^*) \perp_B M = B \perp_B M = M$. This proves that the functors $E \perp -$ and $E^* \perp_B -$ are inverse to each other, which yields the assertion.

(7.50.) Corollary Let (R, σ) be a noetherian torsion couple and E, F a pair of σ-progenerators. Let $B = \operatorname{End}_R(E)$ and assume that F is also a (left) B-module, then $S = \operatorname{Hom}_B(E, F)$ is a σ-progenerator.

Proof From the equivalence in the previous Proposition, it follows that

$$S = \operatorname{Hom}_B(E, F) = [E^* \perp_B E, E^* \perp_B F] = [R, E^* \perp_B F] = E^* \perp_B F.$$

In particular, S is σ-closed. As $S_p = (E^* \perp_B F)_p = E_p^* \otimes F_p$ for all $p \in \mathbf{K}(\sigma)$ and since we know that the latter is free of finite rank over R_p, by a related result in [AG1], clearly S is σ-quasiprojective. To finish the proof, it thus suffices to show that $E^* \otimes_B F$ is σ-finitely generated. But, this is a homomorphic image of $E^* \otimes F$ (tensorproduct over R!), so it will be sufficient to verify whether $E^* \otimes F$ (or $E^* \perp F$) is σ-finitely generated, and this follows from (7.31.)!

(7.51.) Corollary Assume (X, Y) is a noetherian torsion couple and let E, F be a pair of Y-progenerators. Let $B = \mathbf{End}(E)$ and suppose that F is also a sheaf of left B-modules, then $S = \mathbf{Hom}_B(E, F)$ is a Y-progenerator.

(7.52.) Corollary Let (X, Y) be a noetherian torsion couple and let A resp. B be Y-Azumaya algebras on X, then the following assertions are equivalent:

(7.52.1.) A is Y-similar to B;

(7.52.2.) there is a Y-progenerator E on X such that $A \perp B^{\mathrm{opp}} = \mathbf{End}(E)$.

Proof We may reduce to the affine case, i.e. work with σ-Azumaya algebras A and B over a noetherian torsion couple (R, σ) and argue as in [VV2]. Here it suffices to prove the following : if A is a σ-Azumaya algebra and $A \sim Q_\sigma(R)$, then we may find a σ-progenerator E and an isomorphism $A \cong \mathrm{End}_R(E)$. Now, if A is trivial, there exists a pair of σ-progenerators E and F such that $A \perp \mathrm{End}_R(E) = \mathrm{End}_R(F)$. Let $B = \mathrm{End}_R(E)$, then the canonical map $A \otimes B \to A \perp B = \mathrm{End}_R(F)$ endows F with a B-module structure. Let $S = \mathrm{Hom}_B(E, F)$, then it follows from (7.50.) that S is a σ-progenerator and that $S = E^* \perp_B F$. Hence $S \perp E = F$ and it follows that $\mathrm{End}_R(S) = \mathrm{End}_B(S \perp E) = \mathrm{End}_B(F)$. Mimicking the proof of the similar result for plain Azumaya algebras, one immediately deduces from this and (7.49.) that $\mathrm{End}_B(F) = A$, proving the assertion, cf. [DI]. We leave details to the reader.

(7.53.) Theorem Let $Y' \subset Y$ be generically closed subsets of a separated scheme X, which is locally noetherian with respect to Y (and hence with respect to Y') and such that $\underline{O}_X$ is Y' closed, then there is an exact sequence of abelian groups

$$\mathrm{Pic}(X, Y) \to \mathrm{Pic}(X, Y') \to \mathrm{Bcl}(X; Y, Y') \to \mathrm{Br}(X, Y) \to \mathrm{Br}(X, Y').$$

Proof Since $\underline{O}_X$ is Y'-closed, it follows from (7.34.) and its global version that $\wp(X, Y) \subset \wp(X, Y')$, etc. From the definitions, it follows that the composition of any two consecutive maps in this sequence is the zero map. All equivalence classes will be denoted by square brackets [ ], the reader may verify that this causes no ambiguity here.

Let us first check exactness at Pic(X, Y'). Suppose M is a Y'-invertible sheaf of $\underline{O}_X$-modules with the property that $j([M]) = 0$, then by definition we may find a couple of Y-progenerators E and F such that $Q_{Y'}(M \otimes E) = Q_{Y'}(F)$. It then follows from (7.39.) that M is a Y-progenerator too. But then

$$Q_Y(M^* \otimes M) = \mathbf{End}(M) = Q_Y(\underline{O}_X) = \underline{O}_X.$$

This proves that M is Y-invertible as well, hence that $[M] \in \mathrm{Im}(i)$.

To check exactness at Bcl(X; Y, Y'), proceed as follows. Let $E \in \Gamma$ and assume that $\alpha([E]) = 0$, then $\mathbf{End}(E) \sim \underline{O}_X$ (as $\underline{O}_X$ is also Y-closed!). By (7.52.) we may find $P \in \wp(X, Y)$ such that $\mathbf{End}(E) = \mathbf{End}(P)$ and of course $P \in \wp(X, Y')$ as well. Using (7.50.), we have that

$$P \perp \mathbf{Hom}_B(P, E) = P \perp (P^* \perp_B E) = (P \perp P^*) \perp_B E = B \perp_B E = E$$

(where $- \perp_B -$ denotes $Q_{Y'}(- \otimes_B -)$, and where we have applied the global version of (7.50.).) If we write $F = \mathbf{Hom}_B(P, E)$, then $F \in \wp(X, Y')$ and from $F \perp P = E$ and $\mathbf{End}(P) = \mathbf{End}(E)$, one easily derives that $F \in \mathbf{Pic}(X, Y')$.

Since obviously [E] = [F], this shows the exactness of the sequence at Bcl(X; Y, Y').

Finally, let [A] $\in$ Ker(β), then A = **End**(E) for some Y'-progenerator E, by (7.52.). But then [A] $\in$ Im(α), which finishes the proof.

(7.54.) Example Let U be an open subset of a separated, locally noetherian scheme X such that $\underline{Q}_X$ is U-closed - for example, let X = $\mathbf{A}^2$, affine 2-space over a fixed field k, and U = $\mathbf{A}^2$ - {(0, 0)}. If we denote by Bcl(X, U) the abelian group consisting of all classes of $\underline{Q}_X$-modules E which are locally free of finite rank on U and for which **End**(E) is a sheaf of Azumaya algebras on X (or an "Azumaya algebra", as previously), then the foregoing yields an exact sequence

$$Pic(X) \rightarrow Pic(U) \rightarrow Bcl(X, U) \rightarrow Br(X) \rightarrow Br(U).$$

(7.55.) The foregoing may give the impression that our results do not apply to the case X = Y in general, since we do not assume our schemes to be locally noetherian. However, if we look somewhat closer at the proof of (7.53.), it follows that mainly only the validity of (7.39.) is needed. If (X, Y) is a noetherian torsion couple and if E and G are X-progenerators, i.e. locally free sheaves of finite rank on X, then we claim that for any Y-progenerator F such that $Q_Y(E \otimes F) = Q_Y(G)$, we have that F is locally free of finite rank too, whenever $\underline{Q}_X$ is Y-closed.

Indeed, reducing to the affine case, let M and P be R-progenerators and let N be a σ-progenerator for some idempotent kernel functor σ in R-mod with the property that R is σ-noetherian and σ-closed and with the property that $Q_\sigma(M \otimes N) = M \perp N = Q_\sigma(P)$. Let M_1 be an R-module such that $M \oplus M_1 \cong R^n$ for some positive integer n, then $(M \otimes N) \oplus (M_1 \otimes N) = N^n$, hence $M \otimes N$ is σ-closed as a direct factor of N^n, which is σ-closed. Similarly, P is a direct factor of some free R-module R^m, so P is σ-closed since R is. We thus have $M \otimes N = P$, and from this it follows that N is a progenerator as well. Indeed, for some R-progenerator Q we have $Q \otimes P = R^q$ for some positive integer q, so $(Q \otimes M) \otimes N = R^q$, from which we deduce that N is a progenerator, indeed.

(7.56.) Corollary Let (X, Y) be a noetherian torsion couple and assume that $\underline{Q}_X$ is Y-closed, then there is an exact sequence

$$\text{Pic}(X) \to \text{Pic}(Y) \to \text{Bcl}(X, Y) \to \text{Br}(X) \to \text{Br}(Y).$$

Here $\text{Bcl}(X, Y) = \text{Bcl}(X; X, Y)$ with notations as before. This exact sequence generalizes analogous sequences due to B. Auslander [Au1], S. Yuan [Yu], M. Orzech [Or1], H. Lee - M. Orzech [LO1], ...

Appendix: A K-THEORETIC APPROACH.

(7.57.) In this Section we answer (and generalize) a question raised by H. Lee and M. Orzech in [LO1]. Throughout X denotes a separated scheme, $Y' \subset Y$ are generically closed subsets of X and we assume X to be locally Y-noetherian and $\underline{O}_X$ to be Y'-closed. We want to deduce the exact sequence (7.53.) by K-theoretic means, so some familiarity with the contents of [Ba] is assumed. The contents of this Section is related to [Ve7], which it generalizes.

For each generically closed subset Y of X, we will introduce a new category $\mathbf{A}(Y)$ as follows. The objects of $\mathbf{A}(Y)$ are the Y-Azumaya algebras on X. If $A, B \in \mathbf{A}(Y)$, then we let $\Delta(A, B)$ consist of all (P, u, Q), where P and Q are Y-progenerators and $u : A_P \to B_Q$ is an isomorphism, where for any $A \in \mathbf{A}(Y)$ and any $P \in \wp(X, Y)$, we write A_P for $A \perp \mathbf{End}(P)$ (where $\perp$ is the modified tensor product with respect to Y). With this convention, it is clear that if Q is another Y-progenerator, then $(A_P)_Q = A_{P \perp Q}$. Similarly, we write 1_P for $\mathbf{End}(P)$ and if $u : A \to B$ resp. $f : P \to Q$ is an isomorphism in the category of Y-Azumaya algebras resp. Y-progenerators, then we will write u_P resp. A_f for $u \perp \mathbf{End}(P) = u \perp 1_P$ resp. $A \perp \mathbf{End}(f) : A_P \to A_Q$.

We define a relation $\sim$ in $\Delta(A, B)$ by putting $(P, u, Q) \sim (E, v, F)$ if and only if we may find U, V in $\wp(X, Y)$ and isomorphisms $f : P \perp U \to E \perp V$ resp. $g : Q \perp U \to F \perp V$ such that the following diagram is commutative

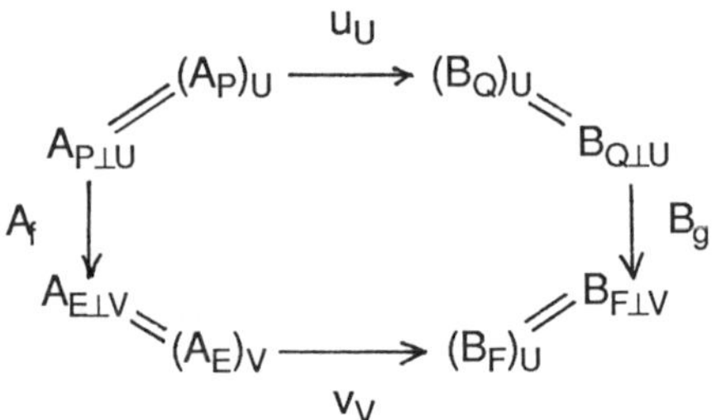

It is clear from the definition that if $A \in \mathbf{A}(Y)$ and $(P, u, Q) \in \Delta(A, A)$, then obviously $(P, u, Q) \sim (A, \mathrm{id}_A, A)$ if and only if we may find $u \in \wp(X, Y)$ and an isomorphism $P \perp U \to Q \perp U$. Finally, let $\mathrm{Hom}_{\mathbf{A}(Y)}(A, B) = \Delta(A, B)/\sim$.

Let us define the composition of "morphisms" as follows : take (P, u, Q) in $\Delta(A, B)$ and (E, v, F) in $\Delta(B, C)$, where $A, B, C \in \mathbf{A}(Y)$, where $P, Q, E, F \in \wp(X, Y)$ and where $u : A_P \to B_Q$ resp. $v : B_E \to C_F$ are isomorphisms. We let

$$(E, v, F) \cdot (P, u, Q) = (P \perp E, v_Q \cdot u_E, F \perp Q) \in \Delta(A, C).$$

If $\phi \in \mathrm{Hom}_{\mathbf{A}(Y)}(A, B)$ is represented by (P, u, Q) and $\psi \in \mathrm{Hom}_{\mathbf{A}(Y)}(B, C)$ is represented by (E, v, F), then we define $\psi \cdot \phi$ to be the class of $(E, v, F) \cdot (P, u, Q)$. One easily checks this to be independent of the chosen representatives of ϕ and ψ and it follows that the category $\mathbf{A}(Y)$ is thus well defined.

The modified tensor product $\perp$ defines a product on $\mathbf{A}(Y)$ in the sense of Bass [Ba]. In order to calculate its K-groups, we need the following Lemma.

(7.58.) Lemma If E and F are Y-progenerators such that there is an isomorphism $u : 1_E \to 1_F$, then we may find a unique class $c \in \mathrm{Pic}(X, Y)$ such that for each I representing c there is an isomorphism $v : I \perp E \to F$, which induces u.

Proof A morphism v as in the statement yields $1_v : 1_{I \perp E} = (1_I)_E = 1_E \to 1_F$. Let $A = 1_E$, then $u : 1_E \to 1_F$ endows F with a structure of a sheaf of left A-modules. As we pointed out in (7.51.), $I = \text{Hom}_A(E, F)$ is a Y-progenerator and $I \perp E = F$. Moreover, it is easily verified that $I \in \underline{\textbf{Pic}}(X, Y)$ and that the class [I] is well determined.

(7.59.) Calculation of $K_0\textbf{A}(Y)$.

Let $\textbf{A} = \textbf{A}(Y)$, then the group $K_0\textbf{A}$ is defined by dividing out the relations

$$<A \perp B> - <A> - <B>,$$

with $A, B \in \textbf{A}$, in the free abelian group generated by the isomorphism classes $<A>$ of objects A in $\textbf{A}$. Now, by definition, any $A, B \in \textbf{A}$ are isomorphic if and only if we may find Y-progenerators P, Q and an isomorphism $u : A_P \to B_Q$, i.e. if and only if $[A] = [B]$ in $Br(X, Y)$. It follows that $K_0\textbf{A}(Y) = Br(X, Y)$.

(7.60.) Calculation of $K_1\textbf{A}(Y)$.

The group $K_1\textbf{A}$ is defined as follows. Let $\Omega\textbf{A}$ be the category consisting of all couples (A, ϕ) where $A \in \textbf{A}$ and ϕ is an $\textbf{A}$-automorphism of A, with obvious morphisms and product $\perp$, then $K_1\textbf{A}$ is the quotient of $K_0\Omega\textbf{A}$ by the relations

$$[(A, \psi\phi)] - [(A, \phi)] - [(A, \psi)],$$

where ϕ and ψ are $\textbf{A}$-automorphisms of A. Now an $\textbf{A}$-automorphism ϕ of A is represented by some (P, u, Q) in $\Delta(A, A)$, where $u : A_P \to A_Q$ is an isomorphism of Y-progenerators. It follows that u induces an isomorphism

$$\omega(u) : u_{A^{opp}} : (A_P)_{A^{opp}} = 1_{A \perp P} \;\to\; (A_Q)_{A^{opp}} = 1_{A \perp Q},$$

hence, by (7.58.) we may find $[l(u)] \in Pic(X, Y)$ and an isomorphism of the form $\tau(u) : l(u) \perp A \perp P \to A \perp Q$ inducing u. Assume that (E, u, F) belongs to $\Delta(A, A)$ also represents f, then we find similarly $[l(v)] \in Pic(X, Y)$ and another isomorphism of the form $\tau(v) : l(v) \perp A \perp E \to A \perp F$ inducing v. Now, $(P, u, Q) \sim (E, v, F)$ yields that there exists U and V in $\wp(X, Y)$ together with a pair of isomorphisms $f : P \perp U \to E \perp V$ resp. $g : Q \perp U \to F \perp V$ such that $\tau(u) \perp U$ and $\tau(v) \perp V$ induce the same isomorphism $1_{A \perp P \perp U} \to 1_{A \perp F \perp V}$, hence $[l(u)] = [l(v)] \in Pic(X, Y)$.

Conversely, to any $l \in \wp(X, Y)$ such that $[l] \in Pic(X, Y)$, we associate the automorphism f of $A \in \mathbf{A}$ represented by $(\underline{O}_X, id, l)$. It is easy to see that f only depends upon $[l]$ and that we thus get a bijection between $K_0 \Omega \mathbf{A}$ and $Pic(X, Y)$. Moreover, if ϕ and ψ are $\mathbf{A}$-automorphisms of A and ϕ resp. ψ correspond to I resp. J, then $\psi\phi$ corresponds to $J \perp I$, hence we actually get an isomorphism between the abelian groups $K_1 \mathbf{A}(Y)$ and $Pic(X, Y)$.

(7.61.) Calculation of $K_1 \Phi F$.

It follows from (7.36.) that our assumptions imply that $\wp(X, Y) \subset \wp(X, Y')$ and this induces an inclusion functor $F : \mathbf{A}(Y) \to \mathbf{A}(Y')$ as one easily verifies. This functor being product preserving, we may construct the abelian group $K_1 \Phi F$ which is defined as follows. Let ΦF denote the category with objects (A, ϕ, B), where A, B belong to $\mathbf{A}(Y)$ and where f belongs to $Hom_{\mathbf{A}(Y')}(A, B)$ is an isomorphism in $\mathbf{A}(Y')$, i.e. we may find P, $Q \in \wp(Y')$ together with an isomorphism $u : A_P \to B_Q$.

The morphisms in ΦF are the $(\alpha, \beta) : (A, \phi, B) \to (A', \phi', B')$, with $\alpha : A \to A'$ resp. $\beta : B \to B'$ are morphisms in $\mathbf{A}(Y)$, which make the following diagram commutative:

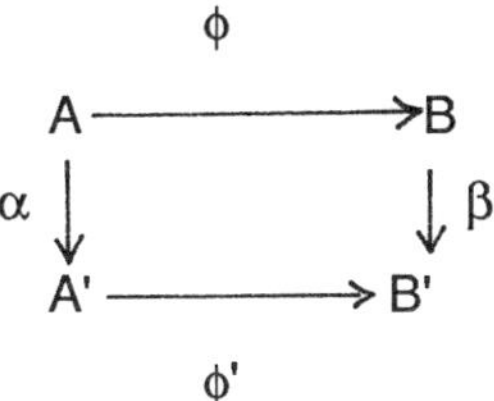

The product in ΦF being defined in the obvious way, we let $K_1\Phi F$ denote the group $K_0\Phi F$ modulo the subgroup generated by the expressions

$$[(A, \psi\phi, C)] - [(A, \phi, B)] - [(B, \psi, C)],$$

where $A, B, C \in \mathbf{A}(Y)$ and where $\phi : A \to B$ resp. $\psi : B \to C$ are isomorphisms in $\mathbf{A}(Y')$.

If $(A, \phi, B) \in \Phi F$, then, with f represented by (P, u, Q), there is an isomorphism

$$\omega(u) = u_{A^{\mathrm{opp}}} : 1_{A\perp P} \to (A^{\mathrm{opp}} \perp B)_Q.$$

Let $C = \mathbf{End}(Q)$ and $S = \mathbf{Hom}_C(Q, A \perp P)$, then, as before, there is an isomorphism $A^{\mathrm{opp}} \perp B \to 1_S$ or, equivalently, an isomorphism

$$\sigma(u) : A \perp P \to S \perp Q$$

in $\wp(X, Y')$, which induces $\omega(u)$. It follows that $S \in \wp(X, Y')$ and from $1_S = A^{\mathrm{opp}} \perp B$, that $1_S \in \wp(X, Y)$. The class of S in $\mathrm{Bcl}(X; Y, Y')$ does not depend upon the choice of representatives (P, u, Q) of ϕ. For if (E, v, F) also represents ϕ, then by definition we may find U and V in $\wp(X, Y')$ together with isomorphisms $f : P \perp U \to E \perp V$ resp. $g : Q \perp U \to F \perp V$. Let $T = \mathbf{Hom}(F, A \perp E)$ with $D = \mathbf{End}(F)$, then f and g induce an isomorphism

$S \perp U = T \perp V$, i.e. the isomorphism classes of S and T define the same element in $Bcl(X; Y, Y')$. With notations as before, it is clear that every S in Γ defines an isomorphism class $(1_S, \phi, \underline{O}_X)$ in ΦF, where $\phi : 1_S \to (\underline{O}_X)_S$ $= 1_S$ is the identity. Applying our construction to $(1_S, \phi, \underline{O}_X)$ permits us to recover S! Since the class of S in the group $Bcl(X; Y, Y')$ does not depend upon the class of A or B in $Br(X, Y)$, it follows that associating $[S]$ $\in Bcl(X; Y, Y')$ to (A, ϕ, B) defines a group homomorphism $K_0\Phi F \to Bcl(X; Y, Y')$, which clearly factorizes through $K_1\Phi F \to Bcl(X; Y, Y')$. The lifting given above yields an inverse to this map, i.e. $K_1\Phi F = Bcl(X; Y, Y')$.

(7.62.) K-theoretic proof of (7.53.).

The inclusion $i : \wp(X, Y) \to \wp(X, Y')$ yields an inclusion $F : \mathbf{A}(Y) \to \mathbf{A}(Y')$ as above. Although i is not cofinal, in general, the functor F is. Indeed, since $A \perp A^{\mathrm{opp}} = 1_A$ for any $A \in \mathbf{A}(Y')$, we obtain that $F(\underline{O}_X) = A \perp A^{\mathrm{opp}}$ in $\mathbf{A}(Y')$, which proves the assertion, since $\underline{O}_X \in \mathbf{A}(Y)$. The long exact sequence for the lower K-groups (the (K_0, K_1)-sequence in [Ba]) applied to F then yields

$$K_1\mathbf{A}(Y) \to K_1\mathbf{A}(Y') \to K_1\Phi F \to K_0\mathbf{A}(Y) \to K_0\mathbf{A}(Y').$$

Hence, applying the above calculation, we obtain

$$Pic(X, Y) \to Pic(X, Y') \to Bcl(X; Y, Y') \to Br(X, Y) \to Br(X, Y'),$$

where one easily verifies the maps to coincide with the ones defined previously. This finishes the proof.

8 INVARIANTS OF KRULL SCHEMES

(8.1.) The purpose of this Section (and the next one) is to present some first applications of the theory developed in the previous Sections. Here we will mainly be concerned with interpreting our results and constructions in the case of Krull domains and Krull schemes. Thus, some of the results in this Section generalize those derived in [VV2] in the more restrictive, noetherian case. Some material, and in particular most of the terminology is drawn from [CF, LO1, LO2]. We start by collecting some results on reflexive sheaves, which strengthen similar statements in [Bar, Ha2, OSS] and which are of interest in their own right. Although some of these results may be derived in a more general set-up, we have chosen for simplicity's sake to consider them mainly in the case of Krull schemes. This will be of great help in the next Section, when we use the concrete interpretation of the previously introduced relative invariants on Krull schemes to study Hecke actions on class groups and divisorial Brauer groups, say.

(8.2.) Assume for a moment (X, Y) to be a noetherian torsion couple in the sense of (7.31.), such that $\underline{O}_X$ is Y-closed. Let E be a sheaf of $\underline{O}_X$-modules on X, then the dual of E is $E^* = \mathbf{Hom}(E, \underline{O}_X)$. We call E <u>reflexive</u> if it is quasicoherent and if the canonical map $E \to E^{**}$ is an isomorphism. Of course, all locally free sheaves of $\underline{O}_X$-modules are reflexive.

Let E be locally of Y-finite type (i.e. E is locally Y-finitely presented, since $\underline{O}_X$ is assumed to be locally Y-noetherian). We claim that E^* is locally of Y-finite type too. Indeed, we may assume X to be affine, i.e. we work over a noetherian torsion couple (R, σ), where R is σ-closed. In this case E is associated to some σ-finitely generated R-module M. Pick a finitely generated R-submodule N of M, where M/N is σ-torsion. Since R is σ-closed, it follows that $M^* = N^*$, where $(-)^* = [- , R]$ as usually. On the other hand, there is a surjective R-linear map $R^n \to N$ for some positive integer n, which by dualizing yields an inclusion $N^* \to (R^n)^* = R^n$. Since R is σ-noetherian, so is R^n and $N^* = M^*$ is σ-finitely generated, as claimed.

In particular, if E is locally of Y-finite type, then so is E^{**}.

It also follows that if E is locally of Y-finite type and reflexive, and if $(X, \underline{O}_X)$ is locally noetherian, then E is coherent. Indeed, $M^* = N^* \subset (R^n)^*$ now yields that M^* (and hence M^{**}) is finitely presented, R being noetherian.

We have the following result:

(8.3.) Lemma Let E be locally of Y-finite type, then the following are equivalent:

(8.3.1.) E is reflexive;

(8.3.2.) $\Gamma(U, E)$ is a reflexive $\Gamma(U, \underline{O}_X)$-module for all open affines U of X.

Proof Let (R, σ) be a noetherian torsion couple, where R is σ-closed and M is a σ-finitely generated R-module, then we claim that $\underline{O}_M{}^* = \underline{O}_{M^*}$. Indeed, if we put $S = R_f$ resp. $N = M_f$ for any $f \in R$, then we have

$$\Gamma(X(f), \underline{O}_M{}^*) = \Gamma(X(f), \mathbf{Hom}(\underline{O}_M, \underline{O}_R)) = [\underline{O}_M|X(f), \underline{O}_R|X(f)] = [\underline{O}_N, \underline{O}_S] = [M_f, N_f].$$

But, as usually, $[M_f, R_f] = [M, R]_f$, since M is σ-finitely presented and R is σ-closed, so

$$\Gamma(X(f), \underline{O}_M{}^*) = (M^*)_f = \Gamma(X(f), \underline{O}_{M^*}).$$

Since the X(f) form a basis for the Zariski topology on Spec(R), this proves $\underline{O}_M{}^* = \underline{O}_{M^*}$, indeed. Moreover, it follows from the foregoing that M^* is σ-finitely generated as well.

From this it follows that the result is true if X is affine. On the other hand, E is reflexive if and only if E|U is a reflexive sheaf of $\underline{O}_X|U$-modules for any open subset U of X, so a straightforward local-global argument finishes the proof in the general case.

Call E <u>locally Y-free</u> if E|Y is locally free on Y. By (7.24.), every Y-progenerator is locally Y-free.

(8.4.) Proposition Let E be locally Y-free and locally of Y-finite presentation, then the following assertions are equivalent:

(8.4.1.) E is Y-closed;

(8.4.2.) E is reflexive.

Proof If E is Y-closed, then it follows that E is a Y-progenerator, hence so are E^* and E^{**} by (7.31.). Let $x \in Y$, then $(E^*)_x = \mathbf{Hom}(E, \underline{O}_X)_x = [E_x, \underline{O}_{X,x}] = E_x^*$, so it is easily checked that the map $E \to E^{**}$ is an isomorphism, since it is isomorphic at each $x \in X$, as E_x is free over $\underline{O}_{X,x}$ and since both E and E^{**} are Y-closed, being Y-progenerators.

Conversely, by (7.27.) it follows that E^* is Y-closed for any E satisfying our assumptions, hence if E is reflexive, then $E = E^{**}$ is Y-closed.

(8.5.) Proposition Let E be locally of Y-finite type, then the following assertions are equivalent:

(8.5.1.) E is reflexive;

(8.5.2.) (locally) E may be included into an exact sequence

$$0 \to E \to F_1 \to F_2,$$

where F_1 and F_2 are (locally) free of finite rank.

Proof Since the question is local, one may reduce the problem to the affine case, so we work over a noetherian torsion couple (R, σ), where R is σ-closed and M is some σ-finitely generated R-module. As we have seen, M^* is then σ-finitely presented, hence we may find an R-linear map $u : N \to M^*$, with σ-torsion kernel and cokernel and such that N is finitely presented, i.e. fitting into an exact sequence

$$R^m \to R^n \to N \to 0,$$

for some positive integers m, n. Dualizing yields an exact sequence

$$0 \to N^* \to (R^n)^* \to (R^m)^*.$$

But, if M is reflexive, then $N^* = M^{**} = M$ and $(R^n)^* = R^n$ resp. $(R^m)^* = R^m$, which proves that (1) implies (2).

Conversely, working locally again, consider an exact sequence

$$0 \to M \to L_1 \to L_2,$$

where L_1 and L_2 are free of finite rank, then first note that M injects into M^{**}, since we have a commutative diagram

$$
\begin{array}{ccc}
M & \longrightarrow & L_1 \\
\downarrow & & \downarrow \\
M^{**} & \longrightarrow & (L_1)^{**}
\end{array}
$$

Let $K = \mathrm{Coker}(M \to L_1)$ and $T = M^{**}/M$, then clearly $T \subset K_1$, hence $M^{**}/M \subset L_2$. But the exact sequence

$$0 \to M \to M^{**} \to T \to 0$$

yields

$$0 \to T^* \to M^{***} \cong M^*,$$

hence $T^* = 0$ and so $T^{**} = 0$. From the commutativity of

$$
\begin{array}{ccc}
T & \longrightarrow & L_2 \\
\downarrow & & \downarrow \\
0 = T^{**} & \longrightarrow & (L_2)^{**},
\end{array}
$$

it finally follows that $T = 0$, i.e. M is reflexive, indeed.

(8.6.) Let us call a scheme X <u>integral</u> if $\Gamma(U, \underline{O}_X)$ is a domain for any open subset U of X. The field of fractions of $\Gamma(U, \underline{O}_X)$ is then denoted by K(X) and called the <u>function field</u> of X. In this case all $\Gamma(U, \underline{O}_X)$ with U open and non-empty have the same field of fractions (= K(X)). Moreover, for any $\emptyset \neq V \subset U$, the restriction morphism $\Gamma(U, \underline{O}_X) \to \Gamma(V, \underline{O}_X)$ is injective. In particular, for any $x \in X$, we have $\underline{O}_{X,x} = \cup \, \Gamma(U, \underline{O}_X)$, where U runs through $V_X(x)$.

From this, one easily deduces that for any open subset U of X we have

$$
\Gamma(U, \underline{O}_X) = \cap \{ O_{X,x} ; x \in U \}.
$$

Let us call a sheaf of $\underline{O}_X$-modules E <u>torsionfree</u> if it is quasicoherent and if for every $x \in X$ the stalk E_x is torsionfree, i.e. if $r \in \underline{O}_{X,x}$ and $e \in E_x$, then $re = 0$ implies $r = 0$ or $e = 0$. Clearly X is integral if and only if $\underline{O}_X$ is torsionfree as a sheaf of $\underline{O}_X$-modules. Any quasicoherent subsheaf of a torsionfree sheaf of $\underline{O}_X$-modules is torsionfree and if X is an integral scheme, then any locally free sheaf of $\underline{O}_X$-modules is torsionfree.

<u>Assume from now on that X is integral</u>. It then follows in particular that a quasicoherent sheaf of $\underline{O}_X$-modules E is torsionfree if and only if $\Gamma(U, E)$ is a torsionfree $\Gamma(U, \underline{O}_X)$-module for all non empty open subsets $U \subset X$.

(8.7.) Lemma Let E be locally of finite presentation, then the following assertions are equivalent:

(8.7.1.) E is torsion free;

(8.7.2.) for every $x \in X$, there exists an open neighborhood U of x and a (locally) free sheaf of $\underline{O}_X$-modules F such that $E|U \subset F|U$.

Proof The second statement implies the first. Conversely, assume E to be torsion free. Since E is locally of finite presentation, we find for any x in X that E_x is finitely generated and that $E_x \otimes K(X) = K(X)^r$ for some positive integer r. Since E_x is torsionfree, $E_x \subset E_x \otimes K(X)$, so up to multiplying by some $h \in \underline{O}_{X,x}$, we thus find an embedding $E_x \subset (\underline{O}_{X,x})^r$, which may be extended to some open neighborhood U of x, since E (and $\underline{O}_X{}^r$!) is locally finitely presented.

(8.8.) Corollary A coherent sheaf on a noetherian integral scheme is reflexive if and only if there locally exists an exact sequence of the form

$$0 \to E \to F \to G \to 0,$$

where F is locally free and G is torsionfree.

Proof This follows immediately from (8.5.) and (8.7.). See also [Ha2].

Intermezzo   Schemes of global dimension at most 2.

(8.9.) Even if $\underline{O}_X$ is not Y-closed, the results above, in particular (8.4.) remain valid, albeit in a slightly modified form. Let us work locally for a moment: assume M to be a σ-progenerator, where σ is an idempotent kernel functor in R-mod, such that R is σ-noetherian. Since $Q_\sigma(R)$ is obviously a σ-progenerator, clearly so is $M^* =: [M, Q_\sigma(R)]$, by (7.31.), since R is assumed to be σ-noetherian. It follows that M^{**} is also a σ-progenerator. For any $p \in \mathbf{K}(\sigma)$, we have $(M^{**})_p = [M^*, Q_\sigma(R)]_p = [(M^*)_p, Q_\sigma(R)_p] = [M_p^*, R_p] = M_p^{**}$, where the latter duals are taken in R_p-mod. Since M is a σ-progenerator, M_p is a free R_p-module of finite rank, hence $M_p \cong (M^{**})_p$. So the canonical morphism $M \to M^{**}$ is isomorphic, since M^{**} is σ-closed by (7.27.) and (7.31.) again.

It thus follows that a sheaf of $\underline{O}_X$-modules E on X, which is locally of Y-finite presentation and locally Y-free for some noetherian torsion couple (X, Y), then E is Y-closed if and only if it is reflexive with respect to $Q_Y(\underline{O}_X)$.

(8.10.) With notations as above, assume R is a domain with the property that gl.dim(R) $\leq$ 2 (i.e. $\text{Ext}_R^i(M, N) = 0$ for all $i \geq 3$ and all R-modules M, N). Let M be a σ-progenerator and let $M^* = [M, S]$, where $S = Q_\sigma(R)$ and where it does not matter whether maps are chosen to be R-linear or S-linear, since all R-linear maps between σ-closed R-modules are automatically S-linear.

By the foregoing, M^* is a σ-progenerator, so in particular M^* is σ-finitely presented. Choose an R-linear map $u : N \to M^*$ with σ-torsion kernel and cokernel together with a presentation

$$R^n \to R^m \to N \to 0.$$

Dualizing yields an exact sequence

$$0 \to [N, R] \to R^m \to R^n,$$

from which it follows that [N, R] is a projective R-module (using the fact that gl.dim(R) $\leq$ 2).

Consider the obvious map

$$\mu : [N, R] \to [N, Q_\sigma(R)].$$

Since N is finitely generated, it is easy to see that $\text{Ker}(\mu)$ and $\text{Coker}(\mu)$ are σ-torsion, hence μ yields an isomorphism $Q_\sigma([N, R]) = Q_\sigma([N, Q_\sigma(R)])$.

On the other hand, $[N, Q_\sigma(R)] = [M^*, Q_\sigma(R)] = M^{**}$, which is σ-closed, so $Q_\sigma([N, R]) = M$.

As $[N, R]$ is a projective R-module and as Q_σ commutes with direct sums, it follows that M is a projective S-module, hence so are M^* and $M \otimes_S M^*$. But then $M \otimes_S M^*$ is σ-closed, as a direct summand of a free S-module, so $M \otimes M^* = Q_\sigma(M \otimes_S M^*) = Q_\sigma(M \otimes_R M^*) = End_R(M) = End_S(M)$, from which we deduce that M is finitely generated as an S-module.

(8.11.) Assume X is a scheme (integral, as always) with $gl.dim(X) \leq 2$, i.e. X may be covered by (or possesses a basis consisting of) open affines $Spec(R)$, where R is a domain with $gl.dim(R) \leq 2$. Let X be locally Y-noetherian with respect to some generically closed subset Y and assume E to be a Y-progenerator. The foregoing implies that E is locally (on open affines) of the form $E|U = \underline{O}_M$, where M is a finitely generated projective $\Gamma(U \cap Y, \underline{O}_X)$-module. In other words, with the terminology of (7.21.), it follows that E is then locally projective of finite type over $Q_Y(\underline{O}_X)$.

As a matter of fact, it now follows that there is a bijective correspondence between Y-progenerators and locally projective sheaves of $Q_Y(\underline{O}_X)$-modules of finite type on X. In particular, a Y-invertible Y-closed sheaf of $\underline{O}_X$-modules corresponds bijectively to an invertible sheaf of $Q_Y(\underline{O}_X)$-modules on X, hence $Pic(X, Y) = Pic(X, Q_Y(\underline{O}_X))$.

One proves similarly that there is a bijective correspondence between Y-Azumaya algebras and locally separable sheaves of $Q_Y(\underline{O}_X)$-algebras on X, in the sense of [Au1], i.e. a central sheaf of $Q_Y(\underline{O}_X)$-algebras A is locally separable, if it is locally projective and localy finitely presented, and if it has the property that the obvious map $A \otimes A^{opp} \to \mathbf{End}(A)$ (tensor product over $Q_Y(\underline{O}_X)$) is an isomorphism. Equivalently, a locallly finitely presented sheaf of $Q_Y(\underline{O}_X)$-algebras A on X, such that A_x is an Azumaya algebra over $Q_Y(\underline{O}_X)_x$ for all $x \in X$. Two of these locally separable sheaves of $Q_Y(\underline{O}_X)$-algebras A resp. B are said to be <u>similar</u> (A ~ B) if there are faithful locally projective and locally finitely presented (or generated) sheaves of $Q_Y(\underline{O}_X)$-modules E, F together with an isomorphism

$$A \otimes \mathbf{End}(A) \cong B \otimes \mathbf{End}(F)$$

(tensor product over $Q_Y(\underline{O}_X)$). As pointed out before, the set of equivalence classes for this relation ~ yields an obvious abelian group $Br(X, Q_Y(\underline{O}_X))$, the Brauer group of the ringed space $(X, Q_Y(\underline{O}_X))$.

The foregoing thus yields:

(8.12.) Proposition For any integral scheme X with gl.dim(X) ≤ 2, there are isomorphisms Pic(X, Y) = Pic(X, $Q_Y(\underline{O}_X)$) resp. Br(X, Y) = Br(X, $Q_Y(\underline{O}_X)$).

(8.13.) Example Let R be a domain with gl.dim(R) ≤ 2 and I a finitely generated ideal of R, then there exists an isomorphism Br(X(I)) = Br(Spec(R), $\underline{O}_S$), where S = $Q_I(R)$ = $\Gamma(X(I), \underline{O}_R)$ and one easily checks that Br(Spec(R), $\underline{O}_S$) = Br(S) = Br($Q_I(R)$). Similarly, Pic(X(I)) = Pic($Q_I(R)$). One should compare this to [VV2, (IV.2.15.)]. One may apply this e.g. to prove that for any U $\subset$ **A**^{2} (over an algebraically closed field k!), we have Pic(U) = Pic($\Gamma(U, \underline{O}_X)$) resp. Br(U) = Br($\Gamma(U, \underline{O}_X)$) (cf. [VV2, (IV.2.16.)]).

(8.14.) Note If one says that X possesses Y-gl.dim(X) ≤ 2 if gl.dim($\Gamma(U \cap Y, \underline{O}_X)$) ≤ 2 for all U in an open affine covering (or basis) of X, then the foregoing results remain valid (with a much easier proof, which we leave as an exercise to the reader). In particular, one then recovers exactly [VV2, (IV.2.15.)], if one applies (8.12.) to an open subset

$X(I)$ of $X = \mathrm{Spec}(R)$ such that $Q_I(R)$ is a (noetherian) domain, with $\mathrm{gl.dim}(Q_I(R)) \leq 2$.

(End of Intermezzo...)

(8.15.) Let us now specialize to so-called Krull schemes. We need some preparations first. Let R be an arbitrary commutative ring and denote by $\mathrm{Spec}^{(n)}(R)$ the set of all prime ideals p of R with $\mathrm{ht}(p) \leq n$. It is then clear that $\mathrm{Spec}^{(n)}(R)$ is a generically closed subset of $\mathrm{Spec}(R)$. We will usually say that R is n-<u>noetherian</u> if it is $\mathrm{Spec}^{(n)}(R)$-noetherian. In this Section, similar expressions, such as n-closed, n-finitely generated, ... will be used as well. We write σ_n for $\sigma_{\mathrm{Spec}^{(n)}(R)}$, hence R is n-closed if and only if it is σ_n-closed, etc. The localization functor at σ_n will be denoted by Q_n.

(8.16.) Example Let R be a domain satisfying the following conditions :
(8.16.1.) R_p is noetherian for all $p \in \mathrm{Spec}^{(1)}(R)$;
(8.16.2.) every $0 \neq x \in R$ is contained in a finite number only of ideals in $\mathrm{Spec}^{(1)}(R)$.
We then claim that R is 1-noetherian. Indeed, let p be a nonzero ideal in $\mathrm{Spec}^{(1)}(R)$. Pick $0 \neq x \in p$ and let $q_1, ..., q_n$ be the other height 1 prime ideals containing x. Choose $y \in p$ with $y \notin q_1 \cup ... \cup q_n$ and $z_1, ..., z_n \in p$ with $\Sigma_i R_p z_i = p R_p$, then $Q_1(p) = Q_1(Rx + Ry + \Sigma_i Rz_i)$. From this it follows that p is 1-finitely generated. On the other hand, if $0 \neq q \in \mathrm{Spec}^{(1)}(R)$, then $q \in \mathbf{L}(\sigma_1)$ i.e. $Q_1(q) = Q_1(R)$, hence q is certainly 1-finitely generated.
It then follows that all prime ideals of R are 1-finitely generated. The result now follows from:

(8.17.) Lemma Let Y be a generically closed subset of $\mathrm{Spec}(R)$, then R is Y-noetherian if and only if every $p \in \mathrm{Spec}(R)$ is Y-finitely generated.
Proof One implication is obvious, so assume every $p \in \mathrm{Spec}(R)$ to be Y-finitely generated. If R is not Y-noetherian, choose an ideal I of R which is

maximal with respect to the property of not being Y-finitely generated. We claim that I is prime, thus yielding a contradiction and finishing the proof. Indeed, otherwise there exist $b, c \in R$ with $b, c \notin I$ and $bc \in I$. It follows that $I + Rb$ and $(I : Rb)$ properly contain I, hence they are both Y-finitely generated. Pick a finitely generated ideal $J' \subset (I : Rb)$ such that $(I : Rb)/J$ is Y-torsion and some finitely generated $J'' \subset I$ such that $I + Rb/J'' + Rb$ is Y-torsion. Let $i \in I$, then $Ki \subset J'' + Rb$ for some $K \in L(Y)$, hence for all k in K we have $ki = j'' + rb$ for some $j'' \in J''$ and some $r \in R$. It follows that $rb = ki - j'' \in I$, so $r \in (I : rb)$ and there exists $L \in L(Y)$ with $Lr \subset J'$. We thus find that $L(ki - j'') \subset J'b$ and so $Lki \subset J'' + J'b$. Denote by $(\,-\,)'$ classes modulo $J'' + J'b$ in I, then $(ki)'$ belongs to $\sigma_Y(I/J'' + J'b)$ and $i' \in \sigma_Y(I/J'' + J'b)$. Since $J'' + J'b$ is finitely generated, this proves the assertion.

The following result will be needed below:

(8.18.) Lemma With notations as before, $K(\sigma_1) = \text{Spec}^{(1)}(R)$.

Proof This has already been pointed out in [Ve3]. As R is commutative, for any idempotent kernel functor σ in R-mod, we have $p \in K(\sigma)$ if and only if $\sigma(R/p) = 0$, for any $p \in \text{Spec}(R)$. Let $q \in \text{Spec}^{(1)}(R)$ and assume that $ht(q) > 1$, so $\sigma_1(R/p) = 0$, i.e. $\sigma_{R\text{-}p}(R/q) = 0$ for some $p \in \text{Spec}^{(1)}(R)$. This means that if $sr \in q$ for some $s \notin p$, then r belongs to q. But, as q does not have height 1, we may pick some s in $q - p$ and applying the foregoing to s yields $q = R$, contradiction.

Conversely, if p has height one, then clearly $p \in K(\sigma_1)$ as $\sigma_{R\text{-}p}(R/p) = 0$. This finishes the proof.

(8.19.) Recall that a domain R is said to be a <u>Krull domain</u> if the following conditions are satisfied:

(8.19.1.) R_p is a discrete valuation ring for all $p \in \text{Spec}^{(1)}(R)$;

(8.19.2.) $R = \cup\{R_p; p \in \text{Spec}^{(1)}(R)\}$;

(8.19.3.) for every $0 \neq x \in R$, there exists only a finite number of $p \in \mathrm{Spec}^{(1)}(R)$, containing x.

From (8.16.) it follows that a Krull domain is 1-noetherian and (8.19.2.) yields that R is $\mathrm{Spec}^{(1)}(R)$-closed, as one easily verifies, using (8.18.). Call a ring R n-regular if R_p is regular for all $p \in \mathrm{Spec}^{(n)}(R)$, then obviously any Krull domain is 1-regular.

One may prove that a domain R is a Krull domain if and only if it is 1-noetherian, 1-closed and 1-regular. One has to prove that if I is an ideal in a Y-noetherian ring R, for some generically closed subset Y of $\mathrm{Spec}(R)$, then the set of prime ideals in Y which are minimal over I is finite. To prove this, one needs some results on relative primary decomposition, see [CF, Ve8] for example.

(8.20.) Let $(X, \underline{O}_X)$ (or X!) be a scheme, then we denote by $X^{(n)}$ the set of all points $x \in X$ with $\dim(\underline{O}_{X,x}) \leq n$. Of course $X^{(n)}$ is generically closed and one easily sees that $\mathrm{Spec}(R)^{(n)} = \mathrm{Spec}^{(n)}(R)$ for any domain R. Instead of speaking of (locally) $X^{(n)}$-noetherian, we will usually just say that X is <u>(locally) n-noetherian</u>.

Following [LO1], we call $(X, \underline{O}_X)$ a <u>Krull scheme</u> if it is separated, quasicompact and integral and if the following conditions are satisfied:

(8.20.1.) $\underline{O}_{X,x}$ is a discrete valuation ring in K(X) for all $x \in X^{(1)}$; denote by v_x the corresponding valuation;

(8.20.2.) for any $0 \neq f \in K(X)$, we have $v_x(f) = 0$ for almost all $x \in X^{(1)}$;

(8.20.3.) if U is an open subset of X, $0 \neq f \in K(X)$ and $v_x(f) \geq 0$ for all $x \in U \cap X^{(1)}$, then $f \in \underline{O}_X(U)$.

Clearly, $\mathrm{Spec}(R)$ is a Krull scheme, whenever R is a Krull domain. Moreover, we also have:

(8.21.) Lemma An integral separated scheme $(X, \underline{O}_X)$ is a Krull scheme if and only if it is quasicompact and if $\Gamma(U, \underline{O}_X)$ is a Krull domain for all affine open subsets U of X.

Proof Let X be a Krull scheme, then it is of the form $U = \mathrm{Spec}(R)$ for some domain $R = \Gamma(U, \underline{O}_X)$, which by definition has the property that R_p is a discrete valuation ring for all $p \in \mathrm{Spec}^{(1)}(R) = X^{(1)} \cap \mathrm{Spec}(R)$. Moreover, to any such p there corresponds a (discrete) valuation v_p and (8.20.2.), (8.20.3.) say exactly that (8.19.3.) and (8.19.2.) hold.

Conversely, if U is quasicompact and $\Gamma(U, \underline{O}_X)$ is a Krull domain for all open affines U of X, then obviously (8.20.1.) holds. Next, let $0 \neq f \in K(X)$ and cover X by a finite number of open affines $\mathrm{Spec}(R)$, where each R is a Krull domain. Since for all of these we have $v_p(f) = 0$ for almost all p in $\mathrm{Spec}^{(1)}(R) = X^{(1)} \cap \mathrm{Spec}(R)$, (8.20.2.) follows, since only a finite number of open sets are used. To prove (8.20.3.), cover U by some open affines $\{U_a = \mathrm{Spec}(R_a); a \in A\}$ and let $0 \neq f \in K(X)$ be such that $v_x(f) \geq 0$ for all x in $U \cap X^{(1)}$, then this also holds for all $x \in U_a \cap X^{(1)} = \mathrm{Spec}^{(1)}(R_a)$, i.e. f belongs to $\underline{O}_X(U_a)$. But $\underline{O}_X(U) = \cap\{\underline{O}_X(U_a); a \in A\}$, since X is integral, hence $f \in \underline{O}_X(U)$, indeed.

(8.22.) It follows from the foregoing that a Krull scheme is 1-noetherian and that $\underline{O}_X$ is 1-closed, where we use for schemes the same type of conventions as described for rings and modules in (8.15.). Conversely, one may prove that an integral scheme X is a Krull scheme if and only if it is 1-noetherian and 1-regular (in the obvious sense) and if $\underline{O}_X$ is 1-closed. In particular, the torsion couple $(X, X^{(1)})$ then satisfies the requirements specified at the beginning of this Section, which allows us to study reflexive sheaves on Krull schemes.

(8.23.) Let R be a Krull domain with field of fractions K. We call an R-module M <u>divisorial</u> if it is torsionfree and if within $K \otimes_R M$ we have

$$M = \cap\{M_p; p \in \mathrm{Spec}^{(1)}(R)\}.$$

The <u>rank</u> of an R-module M is defined as $\dim_K(K \otimes_R M)$. A torsionfree R-module M is said to be an R-<u>lattice</u> if it is of finite rank (i.e. if it spans a finite dimensional vector space over K) and if there exists a finitely generated R-module F with the property $M \subset F \subset K \otimes_R M$. It then clearly follows that $sF \subset M$ for some $0 \neq s \in R$.

(8.24.) Proposition If R be a Krull domain and M a torsionfree R-module, then

(8.24.1.) M is divisorial if and only if it is 1-closed;

(8.24.2.) M is a lattice if and only if it is 1-finitely generated.

Proof The first statement is trivial.

Assume M is a lattice, then it is contained in a finitely generated R-submodule F of $K \otimes_R M$, where K denotes the field of fractions of R. Since R is 1-noetherian by (8.13.), it follows that F is also 1-noetherian, hence by (5.27.1.), that M is 1-finitely generated.

Conversely, let M be 1-finitely generated, then it suffices to find some finitely generated R-module F with $M \subset F \subset K \otimes_R M$. Since M is 1-finitely generated, we may find some $\{x_1, ..., x_m\} \subset M$ such that the module $M/\Sigma_i R x_i$ is 1-torsion. Choose a basis $\{y_1, ..., y_n\} \subset M$ for $K \otimes_R M$, then for some $0 \neq s \in R$, we have $s x_i \in \Sigma_j R y_j$ for all $1 \leq i \leq n$, so $\Sigma_i R x_i \subset \Sigma_j R(s^{-1} y_j) = F$. As $M_p = (\Sigma_i R x_i)_p$ for all p in $\mathrm{Spec}^{(1)}(R)$, we find $M \subset \cap M_p \subset \cap F_p$, where p runs through the height 1 primes of R. To finish the proof, it thus suffices to verify that

$$F = \cap\{F_p; p \in \mathrm{Spec}^{(1)}(R)\},$$

i.e. that F is 1-closed, but this is trivial since F is free of finite rank over R.

(8.25.) Proposition Let R be a Krull domain, then there is a one-to-one correspondence between

(8.25.1.) divisorial R-lattices;

(8.25.2.) 1-progenerators;

(8.25.3.) reflexive R-lattices.

Proof Let M be a divisorial R-lattice, then it is torsionfree, hence 1-closed and finitely generated by (8.24.). Moreover, M_p is finitely generated and torsionfree over R_p, hence it is free over it, as R_p is a discrete valuation ring, for all p in $\mathrm{Spec}^{(1)}(R)$. This proves that M is a 1-progenerator.

Next, assume M is a 1-progenerator. As R is 1-noetherian and M is 1-finitely presented, one proves as in (7.31.) that $M^* = [M, R]$ is 1-finitely presented as well, hence in particular $(M^{**})_p = [(M^*)_p, R_p]$ for all p in $\mathrm{Spec}^{(1)}(R)$. For the same reason $(M^*)_p = (M_p)^*$, the latter dual being taken in R_p-mod, so $(M^{**})_p = (M_p)^{**} = M_p$, hence in particular $Q_1(M^{**}) = Q_1(M) = M$. Finally, one may derive from (7.27.) that M^{**} is 1-closed, hence M is reflexive indeed. But then M is also torsionfree, so it follows from (8.24.) that M is an R-lattice.

Finally, let M be a reflexive R-lattice, then M is 1-finitely presented and M_p is a finitely generated torsionfree R_p-module, for all $p \in \mathrm{Spec}^{(1)}(R)$, hence M_p is free, since R_p is a discrete valuation ring. It then follows that M is 1-quasiprojective, so from (7.27.) it follows that M^* and M^{**} are 1-closed (since R is 1-closed). But $M = M^{**}$ by assumption, so M is 1-closed, i.e. divisorial, as it is also torsionfree.

(8.26.) Let $(X, \underline{O}_X)$ be a Krull scheme, then a torsionfree sheaf of $\underline{O}_X$-modules E is said to be <u>divisorial</u> if for any open subset U of X we have

$$\Gamma(U, E) = \cap\{E_x ; x \in X^{(1)} \cap U\},$$

the intersection being taken within $K(X) \otimes E$ (actually, more precisely, within $K(X) \otimes \Gamma(U, E)$). We call E an $\underline{O}_X$-<u>lattice</u> if it is torsionfree and if

$E(U)$ is an $\underline{O}_X(U)$-lattice for any affine open subset U of X. It then easily follows that any (divisorial) lattice over an affine Krull scheme $Spec(R)$ is of the form $\underline{O}_M$ for some (divisorial) R-lattice M.

From (8.25.) one easily deduces:

(8.27.) Proposition Let $(X, \underline{O}_X)$ be a Krull scheme, then there is a one-to-one correspondence between:

(8.27.1.) divisorial $\underline{O}_X$-lattices;

(8.27.2.) 1-progenerators on X;

(8.27.3.) reflexive $\underline{O}_X$-lattices.

(8.28.) Still assuming $(X, \underline{O}_X)$ to be a Krull scheme, denote by $\perp$ the modified tensor product with respect to $X^{(1)}$, then it is easy to verify that for any pair of torsionfree sheaves of $\underline{O}_X$-modules E, F we have

$$\Gamma(U, E \perp F) = \cap\{E_x \otimes F_x; x \in X^{(1)} \cap U\},$$

the tensorproduct being taken within $K(X) \otimes E \otimes F$. Of course, $E \perp F$ is divisorial and it is an $\underline{O}_X$-lattice if E and F are.

Let E, F be two divisorial $\underline{O}_X$-lattice, then the following properties are direct consequences of corresponding results in the foregoing Sections:

(8.28.1.) if E and F are sheaves of $\underline{A}$-modules for some quasicoherent sheaf of $\underline{O}_X$-algebras $\underline{A}$ on X, then $\mathbf{Hom}_{\underline{A}}(E, F)$ is a divisorial $\underline{O}_X$-lattice and for all $x \in X$ we have

$$\mathbf{Hom}_{\underline{A}}(E, F)_x = [E_x, F_x],$$

the latter morphism set being taken in $\underline{A}_x$-mod; this applies in particular to $\underline{A} = \underline{O}_X$;

(8.28.2.) there are canonical isomorphisms

$$\mathbf{End}(E) \perp \mathbf{End}(F) \cong \mathbf{End}(E \perp F)$$

resp.

$$E \perp E^* \cong \mathbf{End}(E).$$

The reader is invited to compare this to similar results in [LO1].

(8.29.) From now on, we will assume throughout $(X, \underline{O}_X)$ to be a Krull scheme and E an $\underline{O}_X$-lattice. The results below may be generalized outside the scope of Krull domains, except of course for the results where we explicitly have to use that the local rings in codimension one are discrete valuation rings.

Recall e.g. from [GD, I.8.2.10.] that for any subset Z of X, we define $\mathrm{codim}(Z, X) = \inf \dim(\underline{O}_{X,z})$, where $z \in Z$ and dim stands for the Krull dimension as usually. It is easy to prove that if Z is an irreducible closed subset of X, then this is the maximal length n of a chain $Z = Z_0 \subset Z_1 \subset ... \subset Z_n$ of distinct closed irreducible subsets of X. It is also clear that for any open subscheme U of X we have $\mathrm{codim}(Z \cap U, U) \geq \mathrm{codim}(Z, X)$.

Let us call E <u>normal</u> (cf. [Bar, Ha2, OSS]) if for all open subsets U of X and every closed subset Y of U with $\mathrm{codim}(Y, U) \geq 2$ the restriction map $E(U) \to E(U - Y)$ is isomorphic.

It is easy to see that E is normal if and only if this condition holds for all affine subsets U of X. Indeed, let U be an arbitrary open subset of X and Y a closed subset of U with $\mathrm{codim}(Y, U) \geq 2$. Cover U with open affines U_i, then for any i, j, we have $\mathrm{codim}(Y \cap U_i, U_i) \geq 2$ and $\mathrm{codim}(Y \cap U_{ij}, U_{ij}) \geq 2$, where $U_{ij} = U_i \cap U_j$ as usually, hence we get a commutative exact diagram

$$0 \rightarrow E(U) \longrightarrow \Pi_i E(U_i) \longrightarrow \Pi_{i,j} E(U_{ij})$$

$$0 \rightarrow E(U - Y) \rightarrow \Pi_i E(U_i - (U_i \cap Y)) \rightarrow \Pi_{i,j} E(U_{ij} - (U_{ij} \cap Y)),$$

where the rightmost vertical maps are isomorphisms by assumption. Hence the map $E(U) \rightarrow E(U - Y)$ is an isomorphism too, proving the assertion.

(8.30.) Proposition Let E be an $\underline{O}_X$-lattice on X, then the following assertions are equivalent:

(8.30.1.) E is reflexive;

(8.30.2.) E is normal.

Proof Both notions are local, so we may reduce everything to the affine case, i.e. we work with a Krull domain R and an R-lattice M. First assume M to be divisorial, then we want to verify whether $M = \Gamma(X(I), \underline{O}_M)$ for all ideals I of R with $\mathrm{codim}(V(I), \mathrm{Spec}(R)) \geq 2$. It follows immediately that $\dim R_p \geq 2$ for all p in V(I), i.e. $I \subset p$ implies $\mathrm{ht}(p) \geq 2$, so $\mathrm{Spec}^{(1)}(R) \subset X(I)$. Now, M being torsionfree, we find

$$M \subset \Gamma(X(I), \underline{O}_M) = \cap_{p \in X(I)} M_p \subset \cap_{\mathrm{ht}(p) \leq 1} M_p = M,$$

the last equality expressing that M is divisorial, hence $M = \Gamma(X(I), \underline{O}_M)$, indeed.

Conversely, assume that $\underline{O}_M$ is a normal $\underline{O}_R$-lattice. With notations as in (4.4.) and with $i : \mathrm{Spec}^{(1)}(R) \subset \mathrm{Spec}(R)$, we have

$$(i_p^{-1}\underline{O}_M)(\mathrm{Spec}^{(1)}(R)) = \lim \Gamma(X(I), \underline{O}_M),$$

where I runs through the ideals of R such that $X(I) \supset \mathrm{Spec}^{(1)}(R)$.

For each of these we have that $V(I) \cap \mathrm{Spec}^{(1)}(R) = \emptyset$, so $\mathrm{codim}(V(I), \mathrm{Spec}(R)) \geq 2$, and $\Gamma(X(I), \underline{Q}_M) = M$ as $\underline{Q}_M$ is normal. So $(i_p^{-1}\underline{Q}_M)(\mathrm{Spec}^{(1)}(R)) = M$. On the other hand, we leave it as an easy verification to the reader to check that the induced $i_p^{-1}\underline{Q}_M$ is actually a sheaf, i.e. $i_p^{-1}\underline{Q}_M = \underline{Q}_M|\mathrm{Spec}^{(1)}(R)$. But then, by (5.24.) it follows that

$$Q_1(M) = \Gamma(\mathrm{Spec}^{(1)}(R), \underline{Q}_M) = M,$$

hence $M = M^{**}$, by (8.25.).

(8.31.) Note If E is a locally finitely presented $\underline{Q}_X$-lattice (which is true for all of the $\underline{Q}_X$-lattices, if X is locally noetherian!), then the second implication in the foregoing result may also be proved as follows (cf. [OSS]). Let $S(E)$ be the set of all $x \in X$ such that E_x is not free over $\underline{Q}_{X,x}$, then $S(E)$ is a closed subset of X. Of course $X^{(1)} \subset X - S(E)$, since E is torsionfree, hence E_x is free over (the discrete valuation ring) $\underline{Q}_{X,x}$ for any $x \in X^{(1)}$. It thus follows that $\mathrm{codim}(S(E), X) \geq 2$, hence $E(U) \cong E(U - S(E))$ for any open subset U of X. On the other hand, E^{**} is reflexive, hence certainly normal, so we also have an isomorphism $E^{**}(U) \cong E^{**}(U - S(E))$. The diagram

$$
\begin{array}{ccc}
E(U) & \longrightarrow & E^{**}(U) \\
\downarrow & & \downarrow \\
E(U - S(E)) & \longrightarrow & E^{**}(U - S(E))
\end{array}
$$

is commutative, hence to prove the assertion, it suffices to verify the fact that $E(U - S(E)) \cong E^{**}(U - S(E))$. But $E|U - S(E)$ is locally free (as it is locally finitely presented and E_x is free for all $x \in U - S(E)$!), so certainly reflexive, hence we obtain isomorphisms $E|U - S(E) = (E|U - S(E))^{**} = E^{**}|U - S(E)$. This finishes the proof.

(8.32.) Corollary [Ha2, (1.6.)] Let E be a coherent sheaf on a locally noetherian Krull scheme $(X, \underline{O}_X)$, then E is reflexive if and only if it is normal and torsionfree.

(8.33.) Let us link the notion of being normal to some concepts stemming from local cohomology. We need some preparations first.

Let Y be a closed subset of X and $U = X - Y$, then for any sheaf of $\underline{O}_X$-modules E on X, there is an exact sequence

$$0 \to \Gamma_Y E \to E \to \Gamma_U E.$$

Indeed, one easily verifies that it suffices to check that the sequence

$$0 \to \Gamma_Y(X, E) \to \Gamma(X, E) \to \Gamma_U(X, E)$$

is exact, the exactness of the analogous sequence for an arbitrary open subset being derived similarly. First note that $\Gamma_Y(X, E)$ obviously injects into $\Gamma(X, E)$. On the other hand, clearly $\Gamma_U(X, E)$ may be identified with $\Gamma(U, E)$, hence the morphism $p : \Gamma(X, E) \to \Gamma_U(X, E)$ is just the restriction map. If $s \in \Gamma(X, E)$, then $p(s) = 0$ if s vanishes on U, i.e. s has support in Y. This proves the assertion.

Note also that p is surjective if E is flabby, so in this case it follows that

$$0 \to \Gamma_Y E \to E \to \Gamma_U E \to 0$$

is exact. This holds in particular if E is injective, so in this case we find that $\Gamma_U E = E/\Gamma_Y E = \Gamma_{X/Y} E$. Of course, we also have that $\Gamma_U E = j_*(E|U)$, where $j : U \to X$ denotes the canonical inclusion. Since $\Gamma_{X/Y} E = j_*(E|U)$ for any injective E and since Γ_U is left exact, one derives that for <u>any</u> sheaf of $\underline{O}_X$-modules E, we have $H^0_{X/Y} E = j_*(E|U)$. We thus obtain for any E an exact sequence

$$0 \to \Gamma_Y E \to E \to j_*(E|X - Y) \to H^1_Y E \to 0.$$

(8.34.) Proposition For any $\underline{O}_X$-lattice E the following assertions are equivalent:

(8.34.1.) E is normal;

(8.34.2.) E is (X - Y)-closed for any closed subset Y of X with $\mathrm{codim}(Y, X) \geq 2$;

(8.34.3.) for any open subset U of X and any closed subset Y of U with $\mathrm{codim}(Y, U) \geq 2$, the canonical inclusion $j : U - Y \subset U$ induces an isomorphism $E|U \cong j_*(E|U - Y)$.

Proof If X is not locally of finite type with respect to X - Y in (2), then as usually, by (X - Y)-closed, we mean that $E = Cl_{X/Y}E$, as in (6.39.). Of course, $\Gamma_Y E = 0$, since E is assumed to be an $\underline{O}_X$-lattice, hence a torsionfree sheaf of $\underline{O}_X$-modules. It follows that (2) just says that $H^1_Y E = 0$ for any closed subset Y of X with $\mathrm{codim}(Y, X) \geq 2$, in view of (6.38.). It is also easily seen that it is sufficient to check that (3) holds for the special case that U = X, in order to verify the equivalence of (2) and (3). But, E is (X - Y)-closed if and only if $H^1_Y E = 0$ (since $\Gamma_Y E = 0$) and due to the remarks above, this is equivalent to $E = j_*(E|X - Y)$.

To prove that (1) implies (2), consider the exact sequence

$$0 \to E \to j_*(E|X - Y) \to H^1_Y F \to 0.$$

Let U be open in X, then i(U) reduces to the map $E(U) \to E(U - (Y \cap U))$, which is bijective, since $\mathrm{codim}(Y \cap U, U) \geq 2$ and E is assumed to be normal. Hence we obtain that $H^1_Y E = 0$.

Finally, let U be an open subset of X and let Y be a closed subset of U with $\mathrm{codim}(Y, U) \geq 2$. Let $j : U - Y \to U$ be the inclusion and assume $E|U = j_*(E|U - Y)$. Taking global sections, we get that the map from $\Gamma(U, E) = \Gamma(U, E|U)$ to $\Gamma(U, j_*(E|U - Y)) = \Gamma(U - Y, E|U - Y) = \Gamma(U - Y, E)$ is an isomorphism, hence (3) implies (1).

This result strengthens [Ha2, (1.6.)].

(8.35.) As before, $(X, \underline{O}_X)$ denotes a Krull scheme. Denote by $\underline{K}_X$ the sheaf of rational functions on X, i.e. $\Gamma(U, \underline{K}_X) = (S_U)^{-1}\Gamma(U, \underline{O}_X)$ for any open subset U of X, where $S_U = \Gamma(U, \underline{O}_X) - \{0\}$, hence $\underline{K}_X$ is the constant sheaf with fibre $K(X)$, the function field of X.
We have an exact sequence

$$1 \to \underline{O}_X{}^* \to \underline{K}_X{}^* \to \underline{K}_X{}^*/\underline{O}_X{}^* \to 1,$$

where (here!) $(-)^*$ denotes the sheaf of units - no ambiguity should arise. Restricting to $Z = X^{(1)}$, we get an exact sequence

$$1 \to \underline{O}_Z{}^* \to \underline{K}_Z{}^* \to \underline{K}_Z{}^*/\underline{O}_Z{}^* \to 1,$$

where $\underline{K}_Z$ is constructed similarly and $\underline{K}_Z = \underline{K}_X|Z$ as one easily checks. By (2.21.) we get an exact sequence of abelian goups

$$1 \to \Gamma(Z, \underline{O}_Z{}^*) \to \Gamma(Z, \underline{K}_Z{}^*) \to \Gamma(Z, \underline{K}_Z{}^*/\underline{O}_Z{}^*) \to H^1(Z, \underline{O}_Z{}^*) \to \dots$$

Clearly $\Gamma(Z, \underline{K}_Z{}^*) = K(X)^*$ and $H^1(Z, \underline{O}_Z{}^*) = \mathrm{Pic}(Z) = \mathrm{Pic}(X, X^{(1)})$. Moreover, $\Gamma(Z, \underline{O}_Z{}^*) = \Gamma(X, \underline{O}_X{}^*)$. Indeed, cover X by open affines $U_i = \mathrm{Spec}(R_i)$, where R_i is a Krull domain. Since X is separated, $U_{ij} = U_i \cap U_j$ is also affine, say $U_{ij} = \mathrm{Spec}(R_{ij})$, where R_{ij} is also Krull domain. Let $Z_i = Z \cap U_i$ resp. $Z_{ij} = Z \cap U_{ij}$, then we get a commutative exact diagram (with obvious maps):

$$0 \to \Gamma(X, \underline{O}_X) \to \Pi_i \Gamma(U_i, \underline{O}_X) \to \Pi_{i,j} \Gamma(U_{ij}, \underline{O}_X)$$

$$0 \to \Gamma(Z, \underline{O}_Z) \to \Pi_i \Gamma(Z_i, \underline{O}_Z) \to \Pi_{i,j} \Gamma(Z_{ij}, \underline{O}_Z).$$

Since the R_i and R_{ij} are Krull domains, clearly the two vertical maps are isomorphisms. Hence $\Gamma(X, \underline{O}_X) \to \Gamma(Z, \underline{O}_Z)$ is an isomorphism as well.

Denote $\Gamma(Z, \underline{K}_Z{}^*/\underline{O}_Z{}^*)$ by Cart(Z) (the "Cartier divisors" on Z!). It is easy to see that any element of Cart(Z) may be biewed as a family $((U_i, f_i); i \in I)$, where the U_i are open subsets of X with $Z \subset \cup_i U_i$ and where $f_i \in K(X)^*$ for each $i \in I$, such that f_i/f_j belongs to $\Gamma(U_{ij} \cap Z, \underline{O}_X)^*$ for all i, j. We may even choose the U_i to be affine, and then $f_i/f_j \in \Gamma(U_{ij}, \underline{O}_X)^*$ for all i, j.

We thus get an exact sequence

$$1 \to \Gamma(X, \underline{O}_X)^* \to K(X)^* \to \text{Cart}(Z) \to \text{Pic}(X, Z) = \text{Pic}(Z, \underline{O}_Z).$$

We claim that $p : \text{Cart}(Z) \to \text{Pic}(Z)$ is surjective.

(8.36.) Lemma Let X be an arbitrary integral scheme and Y a generically closed subset of X such that X is locally of finite type with respect to Y, then any Y-closed Y-invertible sheaf of $\underline{O}_X$-modules E on X may be embedded into $\underline{K}_X$.

Proof Let $U = \text{Spec}(R)$ be affine and $E|U = \underline{O}_M$ for some σ-closed, σ-invertible R-module, where $U \cap Y = \mathbf{K}(\sigma)$, then $M \otimes_R K(X) \cong K(X)$. It follows that $E \otimes \underline{K}_X|U \cong \underline{K}_X|U$ is a constant sheaf with fibre K(X), hence so is $E \otimes \underline{K}_X$, since X is irreducible. But then $E \otimes \underline{K}_X = \underline{K}_X$, up to isomorphism.

Consider the canonical morphism $i : E \to E \otimes \underline{K}_X$; we claim that i is injective, thus finishing the proof. Indeed, look at the affine case again, with notations as before, then (locally) i corresponds to $j : M \to M \otimes_R K$. Let $p \in \mathbf{K}(\sigma)$, then $\text{Ker}(j)_p = \text{Ker}(M_p \to M_p \otimes_R K) = 0$, since $M_p = R_p$, hence $\mathrm{Ker}(j)$ is σ-torsion, i.e. $\mathrm{Ker}(j) = 0$, since $\mathrm{Ker}(j) \subset M$, which is σ-closed, hence σ-torsionfree. This easily yields the result.

(8.37.) Let $D \in \mathrm{Cart}(Z)$, then this D corresponds to a family $((U_i, f_i); i \in I)$, where $Z \subset \cup U_i$, $f_i \in K(X)^*$ and $f_i/f_j \in \Gamma(U_{ij} \cap Z, \underline{O}_X)^*$ for any i, j. We define $\mathbf{L}(D) \subset \underline{K}_Z$ by putting

$$\mathbf{L}(D)|U_i \cap Z = f_i^{-1}\underline{O}_Y|U_i \cap Z.$$

It is easy to see that $\mathbf{L}(D)$ is a well-defined sheaf of $\underline{O}_Z$-modules on Z. We claim that every invertible $E \subset \underline{K}_Z$ is of this form. Indeed, by definition we may find open subsets $\{U_i; i \in I\}$ covering Z, such that for each $i \in I$, there is an isomorphism

$$\phi_i : \underline{O}_Z|U_i \cap Z \cong E|U_i \cap Z.$$

Let $f_i = \phi_i(U_i \cap Z)(1)^{-1}$, then it is easy to see that $((U_i, f_i); i \in I)$ defines a Cartier divisor $D \in \mathrm{Cart}(Z)$ and that $\mathbf{L}(D) = E$. Clearly D is unique as such. Hence:

(8.38.) Lemma There is a one-to-one correspondence between $\mathrm{Cart}(Z)$ and the set of invertible subsheaves of $\underline{O}_Z$-modules of $\underline{K}_Z$ on Z.

(8.39.) Corollary The map $p : \mathrm{Cart}(Z) \to \mathrm{Pic}(Z)$ is surjective.

(8.40.) Define $\mathrm{Div}(X)$ to be the free abelian group over $Z = X^{(1)}$. For every $z \in Z$, the ring $\underline{O}_{X,z}$ is a discrete valuation ring with associated discrete valuation v_z. By definition, for every $f \in K(X)^*$, we have $v_z(f) = 0$ for almost all $z \in Z$. For any such f we may thus construct

$$(f) = \Sigma_z v_z(f)z,$$

the so-called <u>principal divisor</u> on X associated to f. Denote by Prin(X) the set of these, i.e. Prin(X) is the image of the map

$$K(X)^* \to \text{Div}(X) : f \to (f).$$

Since $v_z(fg) = v_z(f) + v_z(g)$ for all $z \in Z$ and any $f, g \in K(X)^*$, it follows that Prin(X) is a subgroup of Div(X). We denote by Cl(X) the quotient Div(X)/Prin(X) and call it the <u>divisor class group</u> of X.

(8.41.) We construct a map $u : \text{Cart}(Z) \to \text{Div}(X)$ as follows. Assume D belongs to Cart(Z) and represent D by some family $((U_i, f_i); i \in I)$, where Z is covered by the U_i and the f_i belong to $K(X)^*$, with $f_i/f_j \in \Gamma(U_{ij} \cap Z, \underline{O}_X)^*$, then for any $z \in Z$ we put $n_z = v_z(f_i)$, where $z \in U_i$. This is well-defined, since if $z \in U_i \cap U_j$, then f_i/f_j belongs to $\Gamma(U_{ij} \cap Z, \underline{O}_X)^* \subset (\underline{O}_{X,z})^*$, hence $v_z(f_i) - v_z(f_j) = v_z(f_i/f_j) = 0$.

It is easy to see that $u(DD') = u(D) + u(D')$, hence we get a commutative exact diagram of abelian groups

$$1 \to \Gamma(X, \underline{O}_X^*) \to K(X)^* \to \text{Cart}(X^{(1)}) \to \text{Pic}(X, X^{(1)}) \to 1$$

$$1 \to \Gamma(X, \underline{O}_X^*) \to K(X)^* \to \text{Div}(X) \qquad \to \text{Cl}(X) \qquad \to 0$$

and from this it follows that there is a map $v : \text{Pic}(X, X^{(1)}) \to \text{Cl}(X)$.

We claim that v is an isomorphism of abelian groups.

Let us associate to any $D = \sum n_z z \in \text{Div}(X)$ an element of Cart(Z) as follows. Let $\text{Supp}(D) = \{z \in Z; n_z \neq 0\}$, then two possibilities occur:

(a) <u>$z \notin \text{Supp}(D)$</u>; we then pick U_z such that $U_z \cap \text{Supp}(D) = \emptyset$ and $f_z = 1$;

(b) $\underline{z \in \text{Supp}(D)}$; let $f_z = (g_z)^{n_z}q$, where g_z is a generator of the maximal ideal of $\underline{O}_{X,z}$ (which is a discrete valuation ring) and q a unit in $\underline{O}_{X,z}$. Let B consist of the (finite number of) $y \in Z$ such that $v_y(f_z) \neq 0$ and choose U_z with the property that $U_z \cap (\text{Supp}(D) \cup B) = \{z\}$.

We leave it as a straightforward exercise to the reader to verify that the family $((U_z, f_z); z \in Z)$ defines an element in Cart(Z) and that this yields a map w from Cl(X) to Pic(X, $X^{(1)}$), which is easily seen to provide an inverse for v. We thus conclude:

(8.42.) Proposition $Cl(X) = Pic(X, X^{(1)})$.

In other words, the class group of X may be described to consist of isomorphism classes [E] of rank 1 divisorial $\underline{O}_X$-lattices and with multiplication given by $[E].[E'] = [(E \otimes E')^{**}]$.

(8.43.) Let us now briefly consider the functorial behaviour of the class group. If $(X, \underline{O}_X)$ and $(Y, \underline{O}_Y)$ are Krull schemes, then we call any affine scheme morphism $u : (X, \underline{O}_X) \rightarrow (Y, \underline{O}_Y)$ a <u>Krull morphism</u> if it induces a morphism of torsion couples $u : (X, X^{(1)}) \rightarrow (Y, Y^{(1)})$. Locally, u is then of the form $f : (R, \sigma_{1,R}) \rightarrow (S, \sigma_{1,S})$, where for any Krull domain T we denote by $\sigma_{1,T}$ the previously defined idempotent kernel functor σ_1 associated to the height 1 prime ideals of T. Since $\mathbf{K}(\sigma_1,T)$ is just $\text{Spec}^{(1)}(T)$, by (8.12.), it thus follows that f satisfies the PDE ("pas d'éclatement") condition (cf. [Fo, Bo]), i.e. for every $q \in \text{Spec}^{(1)}(S)$, we have $\text{ht}(f^{-1}(q)) \leq 1$, which says that $f^{-1}(q)$ belongs to $\text{Spec}^{(1)}(R)$. In other words, Krull morphisms locally satisfy PDE and vice versa. Since it is well known (and easy to verify) that flat or integral morphisms between Krull domains satisfy PDE, this yields several examples of Krull morphisms.

From (7.17.) and (8.42.) it thus follows immediately:

(8.44.) Proposition Every Krull morphism $u : (X, \underline{O}_X) \to (Y, \underline{O}_Y)$ between Krull schemes induces a canonical group homomorphism

$$Cl(u) : Cl(Y) \to Cl(X).$$

The map $Cl(u)$ is given by sending $[E] \in Pic(Y, Y^{(1)}) = Cl(Y)$ to $[Q_{X^{(1)}}(u^*E)]$. On the other hand, since $\underline{O}_X$ is $X^{(1)}$-closed, it also follows from (7.36.) that $Q_{X^{(1)}}(u^*E) = Q_Z(u^*E)$, with $Z = u^{-1}(Y^{(1)})$. Of course, if $v : (Y, \underline{O}_Y) \to (Z, \underline{O}_Z)$ is another Krull morphism, then $Cl(vu) = Cl(u)Cl(v)$, showing that Cl is actually a contravariant functor from the category of Krull schemes (and Krull morphisms) to the category of abelian groups.

(8.45.) Krull morphisms may also be given a more intrinsic characterization. To show this, we need some preparatory results, which strengthen related results due to M. Orzech.

Let (R, σ) and (S, τ) be torsion couples, such that R is σ-closed and S a τ-closed domain. For any ring morphism $f : R \to S$, consider the following properties:

(8.45.1.) $f : (R, \sigma) \to (S, \tau)$ is a morphism of torsion couples;

(8.45.2.) if M is a σ-progenerator, then $Q_\sigma(S \otimes_R M)$ is a τ-progenerator;

(8.45.3.) f makes S into a σ-closed R-module.

We have seen in (7.34.) that (1) implies (2).

On the other hand, it is easy to see that (2) implies (3). Indeed, since R is a σ-progenerator, applying (2) yields that $Q_\sigma(S)$ is a τ-progenerator. Choose $q \in \mathbf{K}(\tau)$, then $Q_\sigma(S)_q$ has to be a free S_q-module of finite type, which necessarily has rank one (tensor $Q_\sigma(S)_q$ and S_q by L, the field of fractions of S!). Denote by i the canonical inclusion $S \subset Q_\sigma(S)$, then essentially applying Nakayama's Lemma yields that the induced injection $i_q : S_q \subset Q_\sigma(S)_q$ is an isomorphism. Since this holds for all q in

$\mathbf{K}(\tau)$, we obtain that $S = Q_\tau(S) = Q_\tau(Q_\sigma(S)) = Q_\sigma(Q_\tau(S)) = Q_\sigma(S)$, hence S is σ-closed indeed.

Finally, we claim that if S is τ-p.i.d., i.e. if S_q is a principal ideal domain for all q belonging to $\mathbf{K}(\tau)$, then (3) implies (1), i.e. all three statements are equivalent.

Indeed, assume first that S is a principal ideal domain, let $q \in \mathbf{K}(\tau)$ and assume that $f^{-1}(q) \notin \mathbf{K}(\sigma)$, then $f^{-1}(q) \not\subset p$ for all $p \in \mathbf{K}(\sigma)$. So, for each of these $f^{-1}(q)S_p = S_p$, where S is viewed as an R-module through f. But then $qS_p = S_p$ as well, hence the inclusion $q \subset S$ yields an isomorphism $Q_\sigma(q) = Q_\sigma(S) = S$. But q being a principal ideal, it is free, hence σ-closed, since S is. This yields $q = S$, a contradiction.

In the general case, let $q \in \mathbf{K}(\tau)$ and $p = f^{-1}(q)$. Denote by $g : R_p \to S_q$ the induced map and let $Q = qS_q$ resp. $P = pR_p$. With notations as in (6.8.), clearly $Q \in \mathbf{K}(\tau(q))$. Moreover, $S_q = Q_{\sigma(p)}(S_q)$, so it is $\sigma(p)$-closed. Since S_q is a principal ideal domain, we may apply the first part of the proof to show that $(R_p, \sigma(p)) \to (S_q, \tau(q))$ is a morphism of torsion couples. In particular, $P = g^{-1}(Q) \in \mathbf{K}(\sigma(p))$, hence we obtain $p \in \mathbf{K}(\sigma)$, by (6.7.). This finishes the proof.

(8.46.) Corollary ([Or1]) Let $R \subset S$ be an inclusion of Krull domains, then the following assertions are equivalent:

(8.46.1.) S/R satisfies PDE;

(8.46.2.) S is a divisorial R-module;

(8.46.3.) $S \perp M$ is a divisorial S-lattice for every divisorial R-lattice M.

Globally, it follows that a morphism $u : (X, \underline{O}_X) \to (Y, \underline{O}_Y)$ of Krull schemes is a Krull morphism if and only if $u_*\underline{O}_X$ is divisorial as a sheaf of $\underline{O}_Y$-modules on Y, cf. Lee and Orzech [LO1, Prop. 3.1.].

(8.47.) Note It follows in particular that if $R \subset S$ are Krull domains such that S is an R-lattice, then S is necessarily a divisorial R-lattice. Indeed, if S is an R-lattice, then it is contained in a finitely generated $F \subset S \otimes_R K$, hence S is certainly integral over R. But then S satisfies PDE and from (8.46.) it follows that S is divisorial indeed.

(8.48.) Let us now briefly consider the relative Brauer group $Br(X, X^{(1)})$ of X, where as before $(X, \underline{O}_X)$ denotes a Krull scheme. A sheaf of $\underline{O}_X$-algebras A is said to be a <u>divisorial Azumaya algebra</u> (over $\underline{O}_X$) (or, sometimes, a sheaf of divisorial Azumaya algebras) if it is a divisorial $\underline{O}_X$-lattice and if the canonical morphism $m : A \otimes A^{opp} \to \mathbf{End}(A)$ induces an isomorphism $A \perp A^{opp} \cong \mathbf{End}(A)$. From (7.41.) and (8.27.) it follows that a divisorial Azumaya algebra over $\underline{O}_X$ is just an $X^{(1)}$-Azumaya algebra on X.

Two divisorial Azumaya algebras A and B are said to be <u>similar</u> if there exist divisorial $\underline{O}_X$-lattices E, F together with an isomorphism

$$A \perp \mathbf{End}(E) \cong B \perp \mathbf{End}(F).$$

We denote this by $A \sim B$. The set of isomorphism classes of divisorial Azumaya algebras over $\underline{O}_X$ modulo the relation $\sim$ will be denoted by $\beta(X)$ or $\beta(X, \underline{O}_X)$ and called the <u>reflexive</u> or <u>divisorial Brauer group</u> of X (or of $(X, \underline{O}_X)$). Its group structure is given by $[A].[B] = [A \perp B]$, $[A]^{-1} = [A^{opp}]$, where $[A]$ denotes the similarity class of the divisorial Azumaya algebra A. It follows from the definitions and (8.28.) that this is well-defined. Moreover, we also deduce from the foregoing that $\beta(X) = Br(X, X^{(1)})$.

From (7.46.) and (8.43.) it follows:

(8.49.) Proposition Every Krull morphism $u : (X, \underline{O}_X) \to (Y, \underline{O}_Y)$ induces a canonical group homomorphism

$$\beta(u) : \beta(Y) \to \beta(X).$$

Of course $\beta(u)$ is given by $\beta(u)([B]) = [Q_{X^{(1)}}(u^*B)]$. Clearly this defines a (contravariant) functor β, since for any other Krull morphism $v : (Y, \underline{O}_Y) \to (Z, \underline{O}_Z)$ we have $\beta(vu) = \beta(u)\beta(v)$.

(8.50.) Let us define a group $Bcl(X)$ (or $Bcl(X, \underline{O}_X)$) as follows. We let Γ be the set of isomorphism classes of divisorial $\underline{O}_X$-lattices E on X such that **End**(E) is locally free. We may define an equivalence relation $\sim$ on Γ by putting $E \sim F$ if and only if there exist locally free sheaves of $\underline{O}_X$-modules of finite rank M and N and an isomorphism $E \perp M \cong F \perp N$. One easily checks that this makes $Bcl(X) = \Gamma/\sim$ into an abelian group with multiplication induced by the modified tensor product $\perp$ and with $[E]^{-1} = [E^*]$, where $[E]$ is the class of E in $Bcl(X)$. Clearly $Bcl(X) = Bcl(X; X, X^{(1)})$, cf. (7.47.). Since $Pic(X, Y) = Pic(X)$, the Picard group of X, and since $Br(X, X) = Br(X)$, the Brauer group of X (cf. [Gr2]), (7.56.) yields the following result, which generalizes related exact sequencxes discovered by B. Auslander [Au1, Au2] and which appears in [LO1]:

(8.51.) Theorem For any Krull scheme $(X, \underline{O}_X)$, there is an exact sequence

$$1 \to Pic(X) \to Cl(X) \to Bcl(X) \to Br(X) \to \beta(X).$$

This sequence is functorial with respect to Krull morphisms, i.e. any Krull morphism of the form $u : (X, \underline{O}_X) \to (Y, \underline{O}_Y)$ induces a commutative exact diagram

$$1 \to \text{Pic}(Y) \to \text{Cl}(Y) \to \text{Bcl}(Y) \to \text{Br}(Y) \to \beta(Y).$$

$$\downarrow \qquad \downarrow \qquad \downarrow \qquad \downarrow \qquad \downarrow$$

$$1 \to \text{Pic}(X) \to \text{Cl}(X) \to \text{Bcl}(X) \to \text{Br}(X) \to \beta(X).$$

(8.52.) Assume that every divisorial $\underline{O}_X$-lattice E for which **End**(E) is locally free is itself locally free, then we claim that $\text{Bcl}(X) = 1$. Indeed, Pick $E \in G$, then $E \perp E^* = \textbf{End}(E)$ is locally free, hence E and E^* are locally free. But then $[E] = [\underline{O}_X]$ in $\text{Bcl}(X)$, since

$$\underline{O}_X \perp \textbf{End}(E) = E \perp E^*,$$

i.e. $\text{Bcl}(X) = 1$, indeed. It thus follows that the map $\text{Br}(X) \to \beta(X)$ is injective.

We claim that the above assumption is satisfied if every Galois extension of $\underline{O}_{X,x}$ is a factorial domain for any $x \in X$, in particular, if for every $x \in X$ the strict henselization $(\underline{O}_{X,x})^{sh}$ of $\underline{O}_{X,x}$ is factorial. Indeed, let E be a divisorial $\underline{O}_X$-lattice on X and suppose **End**(E) is locally free. By (8.28.1.) it follows that $\textbf{End}(E)_x = \text{End}(E_x)$ for any $x \in X$ and since $\underline{O}_{X,x}$ is local, there exists some Galois extension $G(x)$ of $\underline{O}_{X,x}$ such that $G(x) \otimes \text{End}(E_x) \cong M_n(G(x))$, for some positive integer n (the rank of E!). Using the fact that E is $X^{(1)}$-closed and locally $X^{(1)}$-finitely presented, it is easy to verify that $G(x) \otimes \text{End}(E_x) = \text{End}(G(x) \otimes E_x)$, since $G(x)$ is (faithfully) flat over $\underline{O}_{X,x}$, the latter endomorphisms being $G(x)$-linear, i.e. $\text{End}(G(x) \otimes E_x) \cong M_n(G(x))$. Using the idempotents in this matrix ring, it is fairly easy to see that this implies that up to isomorphism $G(x) \otimes E_x = I^n$ for some ideal I of $G(x)$, which is then necessarily divisorial, hence projective since $G(x)$ is factorial. But then, by faithfully flat descent, it follows that E_x is free over

$\underline{O}_{X,x}$. Essentially applying (8.28.1.) again, it follows that E is locally free indeed.

We thus have proved:

(8.53.) Proposition (Lee - Orzech [LO1], Grothendieck [Gr2]) If X is a Krull scheme with the property that every Galois extension of $\underline{O}_{X,x}$ is a factorial domain (in particular if the strict henselization of $\underline{O}_{X,x}$ is factorial!) for every x in X, then the map $Br(X) \to \beta(X)$ is injective.

(8.54.) The relevance of the foregoing result becomes clear, if one notes that $\beta(X)$ always injects into $Br(K(X))$, the Brauer group of the function field $K(X)$ of X. The proof of this may be found in [LO1] and runs essentially along the lines of the classical proof in [AG2] or its relative counterpart in [VV2]. It makes use of the notion of maximal order over a Krull scheme, mimicking some ideas on the subject, developed by R. Fossum. The foregoing Proposition thus gives a sufficient condition to derive an inclusion $Br(X) \subset Br(K(X))$. Mimicking (III. 3.13.) in [VV2] also easily yields that

$$\beta(X) = \cap\{Br(\underline{O}_{X,x}); x \in X^{(1)}\}$$

within $Br(K(X))$.

9   HECKE ACTIONS

(9.1.) In [RS] Roggenkamp and Scott have amply motivated the study of Hecke actions on Picard groups of rings and schemes. One of their main applications dealt with rings of integers in algebraic number fields, where the Picard group of course reduces to the class group. This allows to derive some of Perlis' results [Pe] on class groups of number fields in a much more elegant context, by exploiting to full extent their functorial features. The techniques developed by Roggenkamp and Scott also show that one may fairly easily generalize Perlis' results to arbitrary Dedekind domains. However, their methods do not seem to apply immediately to class groups of arbitrary Krull domains, as in general (in the non-factorial case) the class group and the Picard group of a Krull domain do not coincide.

The main purpose of this Section is to consider the general question, i.e. to study natural Hecke actions on class groups of Krull domains. As we have seen in the foregoing Section, however, class groups are just a particular instance of relative Picard groups. This was our main motivation for formulating and proving the results below in the context of relative invariants, always keeping in mind the particular example of Krull domains.

It appears that we may indeed define a natural Hecke action on relative Picard groups (and relative Brauer groups) which generalizes the "classical" case. Moreover, the functorial aspects of this consruction yield natural restriction and corestriction maps between class groups, which (for rings of integers of algebraic number fields or more general Dedekind domains) coincide with the usual restriction and corestriction. In the general case, it appears that these maps are connected with the norm map defined by Knus and Ojanguren in [KO2]. Since this norm map also works for the relative case, we start this Section by briefly considering norm maps for relative invariants and we show later that these may be recovered from the functorial features of Hecke actions.

A final remark: although this is not strictly necessary, as one may verify essentially by using some techniques due to Yoshida [Yo], we have preferred to restrict to the less technical case of finite groups, which is sufficient for most applications. For completeness' sake, however, we have briefly indicated what happens in the infinite case, but these remarks are not used elsewhere.

(9.2.) Let $R \subset S$ be commutative rings and σ an idempotent kernel functor (of finite type) in R-mod such that R is a σ-noetherian domain. Assume that R and S are σ-closed, then following [CV, VV2] we say that $R \subset S$ is a σ-<u>separable</u> extension or that S is σ-separable over R, if S_p is separable over R_p for all p in $\mathbf{K}(\sigma)$ or equivalently, that S is σ-quasiprojective over $S \perp S$ through the canonical map $\mu : S \perp S \to S$.

On the other hand, S is said to be <u>really</u> σ-<u>separable</u> over R (the terminology is due to S. Caenepeel) if μ makes the algebra S into a projective $S \perp S$-module or, equivalently if there exists some (idempotent) $e \in S \perp S$ with the property that $\mu(e) = 1$ and $e(a \perp 1 - 1 \perp a) = 0$ for all $a \in S$, where $a \perp b$ denotes the canonical image of $a \otimes b \in S \otimes S$ through the localization map $S \otimes S \to S \perp S$.

(9.3.) Proposition If S is σ-separable over R and σ-finitely generated as an R-module, then S is really σ-separable.

Proof If S is σ-finitely generated over R, then S^e is also σ-finitely generated over R, hence S^e is σ-noetherian, as R is still assumed to be σ-noetherian, hence in the exact sequence

$$0 \to J(S) \to S^e \to S \to 0,$$

the R-module $J(S)$ is σ-finitely generated. A fortiori, $J(S)$ is σ-finitely generated over S^e. Hence S is a σ-finitely presented S^e-module. In particular, for any p in $K(\sigma)$ and any σ-closed S^e-module N, we have an isomorphism $(\mathrm{Hom}_{S^e}(S, N))_p = \mathrm{Hom}_{S^e}(S_p, N_p)$. Now, for any $p \in K(\sigma)$, look at the following diagram:

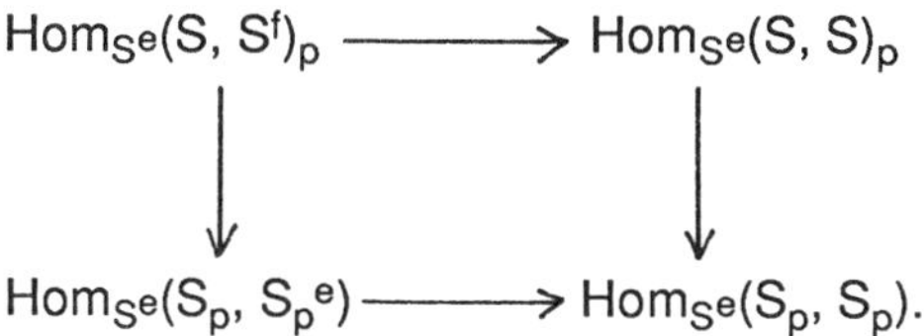

Here $S^f = Q_\sigma(S^e)$; the vertical arrows are isomorphisms by the foregoing and the lower horizontal map is isomorphic since S_p is separable over R_p. It follows that the upper horizontal map is isomorphic too, and as this holds for all $p \in K(\sigma)$, we obtain that μ induces an isomorphism $Q_\sigma(\mathrm{Hom}_{S^e}(S, S^f)) = Q_\sigma(\mathrm{Hom}_{S^e}(S, S))$. But $\mathrm{Hom}_{S^e}(S, S^f)$ and $\mathrm{Hom}_{S^e}(S, S)$ are σ-closed (since S is σ-quasiprojective and both S and S^f are σ-closed), so $\mathrm{Hom}_{S^e}(S, S^f) = \mathrm{Hom}_{S^e}(S, S)$ and S is a direct summand of S^f, hence a projective S^f-module, indeed.

(9.4.) Recall from [VV2] that a commutative R-algebra S is a σ-<u>Galois extension</u> of R if the following conditions are satisfied:

(9.4.1.) S is a σ-separable extension of R and a σ-progenerator;

(9.4.2.) there exists a finite $G \subset \mathrm{Aut}_R(S)$ such that $R = S^G$;

(9.4.3.) Let σ' denote the idempotent kernel functor in S-mod induced by σ, then for every $p \in K(\sigma')$ and every $g \neq 1$ in G, there exists $s \in S$ with $g(s) - s \notin p$.

It has been proved in [VV3] that this is equivalent to the property that for all $p \in K(\sigma)$ the extension S_p of R_p is Galois, with (fixed) group G. From the foregoing it follows that any σ-Galois extension $R \subset S$ is really σ-separable and one may verify that $S \perp S \cong E(S, G)$, where S acts on the first factor in $S \perp S$ and where $E(S, G)$ is a sum of $|G|$ copies of S, the isomorphism being given by sending $a \perp b$ in $S \perp S$ to $\sum_{g \in G} ag(b)e_g$, where the e_g ($g \in G$) form a free basis for $E(S, G)$. This settles a question raised in [VV3].

If R and S are Krull domains and $\sigma = \sigma_1$, the idempotent kernel functor associated with $\mathrm{Spec}^{(1)}(R)$, then we say that S is <u>divisorially Galois</u> over R (with Galois group G) if S is a σ_1-Galois extension with group G over R. In particular, S is then a divisorial R-lattice with $S^G = R$ and S is a Galois extension of R in the sense of [LO2].

(9.5.) Let P be a σ-progenerator, then $\mathrm{End}_R(P) = P \perp P^*$ and we may define the trace $\mathrm{End}_R(P) = P \perp P^* \to R$ to be induced by the morphism $P \otimes P^* \to R : p \otimes \phi \to \phi(p)$. For any R-algebra S which is also a σ-progenerator, this yields as usually a bilinear form $\mathrm{Tr}_{S/R} : S \times S \to R$, which is easily seen to give the usual trace at each $p \in \mathbf{K}(\sigma)$. In particular, $\mathrm{Tr}_{S/R}(x, y) = \mathrm{tr}(xy)$, where $\mathrm{tr} : S \to R$ is obtained by taking the trace of the regular representation $S \to \mathrm{End}_R(S)$. For any finite set $x_1, ..., x_n$ in S, we denote by $D(x_1, ..., x_n)$ the determinant of the $(\mathrm{Tr}_{S/R}(x_i, x_j))_{i, j}$ and by $\delta_\sigma(S/R)$ the localization at σ of the ideal of R generated by the $D(x_1, ..., x_n)$ for all n-tuples $(x_1, ..., x_n)$. We then have as in [KO2]:

(9.6.) Proposition Let S be a commutative R-algebra, then with assumptions as before, the following assertions are equivalent:

(9.6.1.) S is a σ-progenerator and S is σ-separable;

(9.6.2.) for some R-algebra T which is a σ-progenerator and σ-separable, there is an isomorphism of T-algebras $S \perp T \cong T^n$;

(9.6.3.) for some finitely generated σ-faithfully flat R-algebra T, there is an isomorphism of T-algebras $S \perp T \cong T^n$;

(9.6.4.) S is a σ-progenerator over R and the bilinear form $\mathrm{Tr}_{S/R}$ induces an isomorphism $S \cong S^*$;

(9.6.5.) S is σ-finitely generated as an R-algebra and $\delta_\sigma(S/R) = R$.

Proof Let us first check that (4) and (5) are equivalent. Recall from [KO1] that the following properties imply each other:

(i) $\mathrm{Tr}_{S/R}$ is nondegenerate and S is free of finite rank n over R;

(ii) there exist generators $x_1, ..., x_n$ for S such that $D(x_1, ..., x_n)$ is a unit of R.

Now, (4) is equivalent to saying that S is σ-finitely generated over R and that for all $p \in \mathbf{K}(\sigma)$ the map $\mathrm{Tr}_{S/R}$ induces an isomorphism $S_p = S_p{}^*$, together with the fact that S_p is free of finite rank over R_p. The last statement is equivalent to the property that S is σ-finitely generated over R and that for all $p \in \mathbf{K}(\sigma)$, there exist generators $x_1, ..., x_n$ of S_p over R_p such that $D(x_1, ..., x_n)$ is a unit of R_p. Note that this implies that $\delta_\sigma(S/R)_p = R_p$ for all $p \in \mathbf{K}(\sigma)$, hence that $\delta_\sigma(S/R) = R$, indeed. This proves that (4) and (5) are equivalent.

To show that $(1) \Rightarrow (2)$, first note that for any σ-finitely generated σ-quasiprojective R-module P we may define a local rank function

$$\mathrm{rk}_{(R, \sigma)}(P) : \mathbf{K}(\sigma) \to \mathbf{Z}$$

which maps any p in $\mathbf{K}(\sigma)$ to $\mathrm{rk}(P_p)$. But R being a domain by assumption, it follows that $\mathrm{rk}_{(R, \sigma)}$ is necessarily constant. Indeed, if for some $p \in \mathbf{K}(\sigma)$, we have $P_p = (R_p)^{n(p)}$, then $K^{n(p)} = K \otimes_R (R_p)^{n(p)} = K \otimes_R P_p = K \otimes_R P$, i.e. $n(p)$ is independent if $p \in \mathbf{K}(\sigma)$ (K is the field of fractions of R!).

Let us work by induction on the rank of S.

If $\mathrm{rk}_{(R, \sigma)}(S) = 1$, then $S = R$.

Assume that the assertion holds for all morphisms of R-algebras $S \to T$, where S and T are σ-closed, T is a σ-separable s-progenerator over S, with $\mathrm{rk}_{(S, \sigma)}(T)$ constant and $\mathrm{rk}_{(S, \sigma)}(T) < \mathrm{rk}_{(R, \sigma)}(S)$. Obviously, $S \to S \perp S$ is one of these. Indeed, $S \perp S$ is σ-closed and σ-finitely generated over S (which is σ-noetherian, being a s-finitely generated module over the σ-

noetherian domain R!) and for all $p \in \mathbf{K}(\sigma)$ the algebra $S_p \otimes S_p$ is a separable S_p-progenerator. Finally, since now S is of "σ-constant rank" over R, it follows that $rk_{(S,\,\sigma)}(S \perp S)$ is constant as well. Now, since S is σ-separable and σ-finitely generated, there exists an idempotent e in $S \perp S$ such that $(S \perp S)e = S$. Now $S \perp S = (S \perp S)e \oplus (S \perp S)(1 - e)$, where the rank of $(S \perp S)e$ and of $(S \perp S)(1 - e)$ over S is strictly inferior to that of S over R. The induction hypothesis implies that there exist S-algebras T_1, T_2 which are σ-separable and σ-progenerators (over S) and with the property that $(S \perp S)e \perp T_1 = (T_1)^n$ resp. $(S \perp S)(1 - e) \perp T_2 = (T_2)^m$. It follows that $T = T_1 \perp T_2$ does the trick (modified tensorproducts are over S!).

To prove (2) $\Rightarrow$ (3), recall the following from [VV3]. An R-module M is said to be σ-<u>flat</u> if $M \perp$ - is exact in (R, σ)-mod. It is σ-<u>faithfully flat</u> if it is σ-flat and if for any σ-closed module N in R-mod, we have $M \perp N = 0$ if and only if $N = 0$. This is equivalent to the fact that a sequence $N' \to N \to N''$ is exact in (R, σ)-mod if and only if the induced sequence $M \perp N' \to M \perp N \to M \perp N''$ is exact in (R, σ)-mod or still to the property that M is σ-flat and $Q_\sigma(M/pM) \neq 0$ for all $p \in \mathbf{C}(\sigma)$ (the maximal elements in $\mathbf{K}(\sigma)$). In order to apply local-global arguments, use the fact that M is σ-faithfully flat if and only if M_p is faithully flat (over R_p) for all $p \in \mathbf{K}(\sigma)$.

Since a σ-progenerator is of course σ-faithfully flat, (2) $\Rightarrow$ (3) is trivial. On the other hand, recall from [VV3] that if T is σ-finitely generated and σ-faithfully flat, then a σ-closed R-module E is a σ-progenerator if and only if the S-module $T \perp E$ is a σ-progenerator (over S). So, to prove (3) $\Rightarrow$ (4), note that from $S \perp T \cong T^n$ it thus follows that S is a σ-progenerator over R. Moreover, the morphism $S \to S^*$ induced by the trace yields an isomorphism $S \perp T \cong S^* \perp T$ by applying $- \perp T$, hence $S \to S^*$ is an isomorphism.

Finally, (4) $\Rightarrow$ (1) may be proved by a local-global argument. Indeed, to prove that S is σ-separable over R, localize at p, then S_p is projective (hence free) over R_p and Tr induces $S_p \cong (S_p)^*$, hence S_p is separable over R_p for all $p \in \mathbf{K}(\sigma)$. This finishes the proof.

(9.7.) We have already pointed out in the foregoing proof that if S is σ-faithfully flat over R, then an R-module M which is σ-closed is a σ-progenerator over R if and only if $S \perp M$ is a σ-progenerator over S (= a σ'-progenerator). This is only one aspect of descent theory with respect to σ-faithfully flat extensions. In fact, in the relative case, descent theory works as follows.

Let S be a commutative R-algebra and consider $M_1, ..., M_n \in$ R-mod. Define the morphism $e_i : M_1 \otimes ... \otimes M_n \to M_1 \otimes ... \otimes M_{i-1} \otimes S \otimes M_{i+1} \otimes ... \otimes M_n$ by putting $e_i(m_1 \otimes ... \otimes m_n) = m_1 \otimes ... \otimes m_{i-1} \otimes 1 \otimes m_{i+1} \otimes ... \otimes m_n$ and let $\varepsilon_i = Q_\sigma(e_i)$. Let t_{ij} denote the usual "switch-map" and put $\tau_{ij} = Q_\sigma(t_{ij})$:

$$M_1 \perp ... \perp M_i \perp ... \perp M_j \perp ... \perp M_n \to M_1 \perp ... \perp M_j \perp ... \perp M_i \perp ... \perp M_n$$

Finally, given $M'_1, ..., M'_n \in$ R-mod and $f : M_1 \perp ... \perp M_n \to M'_1 \perp ... \perp M'_n$, we define f_i by $f_i = \tau_{i, n+1}(f \perp S)\tau_{i, n+1}$:

$$M_1 \perp ... \perp M_{i-1} \perp S \perp ... \perp M_n \to M'_1 \perp ... \perp M'_{i-1} \perp S \perp ... \perp M'_n$$

The starting point of descent theory may then be formulated as:

(9.8.) Theorem ([CV, VV3]) Let R be σ-closed and S a σ-closed σ-faithfully flat commutative R-algebra. Let M be a σ'-closed S-module and let there be given an $S \perp S$-isomorphism $f : S \perp M \to M \perp S$ with $f_2 = f_3 f_1$. Then there exists a σ-closed R-module N and an S-isomorphism $n : N \perp S \to M$ such that the following diagram of $S \perp S$-isomorphisms is commutative:

$$
\begin{array}{ccc}
 & S \perp n & \\
S \perp N \perp S & \longrightarrow & S \perp M \\
\tau_{12} \downarrow & & \downarrow f \\
N \perp S \perp S & \longrightarrow & M \perp S \\
 & n \perp S &
\end{array}
$$

The pair (N, n) is determined up to isomorphism and one may take

$$
N = \{x \in M; \varepsilon_2(x) = f_1(x)\} = \mathrm{Ker}(\varepsilon_2 - f_1),
$$

and n the map induced by the multiplication. Moreover, if M has the structure of an S-algebra and f is an $S \perp S$-algebra isomorphism, then N may be given the structure of an R-algebra in an essentially unique way, such that the morphism n is an S-algebra isomorphism.

(9.9.) Corollary If S is a σ-progenerator over R and A a σ-closed R-algebra with the property that $A \perp S \cong \mathrm{End}_S(P)$ for some σ-progenerator P over S, then A is a σ-Azumaya algebra. Similarly, a σ-closed R-module M is σ-invertible if and only if $M \perp S$ is σ-invertible as an S-module.

(9.10.) These results will be among the main ingredients of the proof of (9.11.) below. Let us call an R-algebra S σ-<u>étale</u> if it is σ-finitely generated and σ-closed as an R-module and if it satisfies one of the equivalent conditions of (9.6.) It is of course also possible to study σ-étale algebras which are not necessarily σ-finitely generated, but the present notion will do for our purposes. Clearly any σ-Galois extension $R \subset S$ is σ-étale, due to the isomorphism $S \perp S \cong E(S, G)$. In particular, if R and S are Krull domains, then S is σ_1-étale if it is divisorially Galois. If S is some σ_1-étale extension of a Krull domain R, then we call S a <u>divisorially étale lattice</u>.

We are now ready to prove the following result, which generalizes an analogous result in [KO2].

(9.11.) Theorem To any σ-étale extension $R \subset S$, we may associate a functor

$$N^S_R : (S, R)\text{-mod} \rightarrow R\text{-mod},$$

(where (S, R)-mod denotes the category of S-modules and semilinear maps with respect to R-automorphisms of S), with the following properties:

(9.11.1.) $N^S_R(M) = N^S_R(Q_\sigma(M))$ for all $M \in (R, \sigma)$-mod;

(9.11.2.) if $S = R^d$, then up to localizing at σ, the functor N^S_R coincides with the norm defined in [KO2];

(9.11.3.) $N^S_R(S) = R$; $N^S_R(S \otimes_R N) = N \perp ... \perp N$ (d times), if $N \in R$-mod and d is the σ-rank of S over R;

(9.11.4.) $N^S_R(M \otimes_S N) = N^S_R(M) \perp N^S_R(N)$ for all $M, N \in (S, R)$-mod;

(9.11.5.) $N^{S \times T}_R(M \times N) = N^S_R(M) \perp N^T_R(N)$ for all M in (S, R)-mod resp. all N in (T, R)-mod;

(9.11.6) $N^S_R(N^T_S(M)) = N^T_R(M)$ for all $M \in (T, S)$-mod;

(9.11.7.) if S, T, U are R-algebras and $M \in (T, S)$-mod, then $N^T_S(M) \perp U$ $= N^{T'}_{S'}(M \perp U)$, where $S' = S \perp U$, $T' = T \perp U$;

(9.11.8.) if $\lambda_s : S \rightarrow S$ is given by $s' \rightarrow ss'$, then $N^S_R(\lambda_s) = \lambda_{\det(s)}$.

Moreover, N^S_R is completely determined by (1), (2) and (7).

Proof The norm referred to in (2) is defined as follows. Let $S = R^d$ for some positive integer d and let $\{f_i; 1 \leq i \leq d\}$ denote the canonical basis for S over R. Any S-module M is isomorphic to a product $M_1 \times ... \times M_d$ of R-modules $M_i = f_i M$. If $M \in (S, R)$-mod, then one defines $N^S_R(M) = M_1 \otimes ...$ $\otimes M_d$. Let $\tau : S \rightarrow S$ be an R-automorphism, then one may show that there exists a complete set of orthogonal idempotents $e_1, ..., e_r$ of R and permutations $\pi(1), ..., \pi(r)$ of $\{1, ..., d\}$ such that $\tau(e_i f_j) = e_i f_{\pi(i)}(j)$. Moreover, the e_i are uniquely determined, if $\{e_1, \ldots, e_r\}$ is a minimal complete set of idempotents with this property. On the other hand, if the map $f : M \to N$ is τ-semilinear (a morphism in (S, R)-mod!), then with these idempotents one defines an R-linear map

$$\mathbf{N}^S{}_R(\tau, f) \ : \ M_1 \otimes \ldots \otimes M_d \ \to \ N_1 \otimes \ldots \otimes N_d,$$

given by $\mathbf{N}^S{}_R(\tau, f)(e_i m_1 \otimes \ldots \otimes e_i m_d) = f_{k(1)}(e_i m_{k(1)}) \otimes \ldots \otimes f_{k(d)}(e_i m_{k(d)})$, where $\pi(i)(k(j)) = j$ for $j = 1, \ldots, d$ and $f_{k(j)} : e_i M_{k(j)} \to e_i N_j$ is induced by f. One easily verifies this to give a functor

$$\mathbf{N}^S{}_R \ : \ (S, R)\text{-mod} \ \to \ R\text{-mod}.$$

We only give a brief outline of the proof, as it rather closely mimicks that of Theorem 3.2. in [KO2].

One starts from a spliting $\phi : S \perp T \to T^d = X$ of S. For any S-module M, we make $M \perp T$ into an X-module through ϕ, hence $M \perp T = M_1 \times \ldots \times M_d$, and by (1) and (2) we get $N^X{}_T(M \perp T) = \mathbf{N}^X{}_T(M \perp T) = M_1 \perp \ldots \perp M_d = M^\circ$. Just as in loc. cit. one now defines a descent datum for T to R on M°. Looking at the actual construction, it is easy to see that for any $p \in \mathbf{K}(\sigma)$ one has $N^S{}_R(M)_p = \mathbf{N}^{S'}{}_{R'}(M_p)$, where $S' = S_p$ resp. $R' = R_p$ and where $\mathbf{N}^{S'}{}_{R'} : (S', R')\text{-mod} \to R'\text{-mod}$ is the Knus-Ojanguren norm, so, once one has established the unicity of $N^S{}_R$, properties (3), (4), (5), (6) and (8) follow directly through a local-global argument.

Now, the construction of $N^S{}_R$ only depends upon the splitting $\phi : S \perp T \to T^d = X$, so let $\phi_1 : S \perp T_1 \to T_1{}^d$ be another splitting with corresponding functor N_1, then we want to show that $N_1 = N^S{}_R$. But, considering $T \perp T_1$ (which contains both T and T_1!), it is obvious that we only have to look at the cases where $T = T_1$ and $\phi_1 : S \perp T \to T^d$ another T-isomorphism and the case where $T \subset T_1$ together with $\phi_1 = f \perp T_1$.

The second case is trivial, of course, so we just have to look at the first case.

Put $M \perp T = M_1 \times \ldots \times M_d$ resp. $M \perp T = N_1 \times \ldots \times N_d$ depending on whether we consider $M \perp T$ as a $T^d = X$-module through ϕ or ϕ_1. Let $N = N_1 \perp \ldots \perp N_d$, then the identity $1 : M_1 \times \ldots \times M_d \to N_1 \times \ldots \times N_d$ is $\phi^{-1}\phi_1$-semilinear and induces a commutative diagram

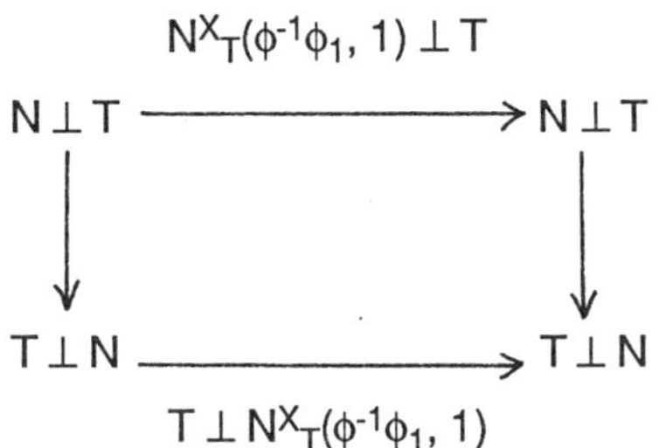

where the vertical maps are the descent data defined by ϕ and ϕ_1. It follows that the restriction of $(\phi^{-1}\phi_1, 1)$ to $N^S{}_R(M)$ yields the desired isomorphism. Since (1), (2) and (7) completely determine the descent datum, this yields the assertion.

(9.12.) Note Consider the special case where $R \subset S$ is a σ-Galois extension with Galois group G. If M is an S-module, denote by M_g ($g \in G$) the g-th conjugate of M, i.e. S acts on M_g through twisting by g. For a fixed g and any $h \in H$ define the morphisms $g^h : M_h \to M_{gh}$ in the obvious way, then g^h is g-semilinear and the g^h induce a g-semilinear map

$$g^\# : \perp_{h \in G} M_h \to \perp_{h \in G} M_h.$$

This yields a "Galois descent datum" $\{g^\#; g \in G\}$ on $\perp M_h$. It is an easy exercise to see that $N^S{}_R(M) = (\perp_{h \in G} M_h)^G$ in this situation. Indeed, by (9.4.)

S_p is a Galois extension of R_p with Galois group G for each $p \in \mathbf{K}(\sigma)$ so by [KO2] we know that

$$N^S{}_R(M)_p = N^{S'}{}_{R'}(M_p) = (\otimes_{h \in G} M_{p,h})^G = ((\perp_{h \in G} M_h)_p)^G,$$

where $R' = R_p$ and $S' = S_p$ as before. Since G is finite, $((\perp_{h \in G} M_h)_p)^G = ((\perp_{h \in G} M_h)^G)_p$, so the assertion follows from the fact that N^G is σ-closed, whenever N is and when G acts trivially on R.

(9.13.) Let us now look at some first examples and applications. We will explicitly consider the case of Krull domains. It has already been proved in (8.46.) that if $R \subset S$ is an extension of Krull domains, then this extension satisfies PDE if and only if S is divisorial over R. In particular, a divisorial S-module is necessarily divisorial, when considered over R.

Now, assume S is a divisorial R-lattice and $\sigma_1 = \inf\{\sigma_{R\text{-}p}; p \in \text{Spec}^{(1)}(R)\}$ as usually. We claim that $Cl(S) = Pic(S, \sigma_1)$ and $\beta(S) = Br(S, \sigma_1)$. Indeed, let $\sigma_{1,S} = \inf\{\sigma_{S\text{-}P}; P \in \text{Spec}^{(1)}(S)\}$ and note that a $\sigma_{1,S}$-closed (= divisorial) S-lattice is also a σ_1-closed (= divisorial) R-lattice. Conversely, if an S-module E is a (σ_1-closed) R-lattice, then it is also a ($\sigma_{1,S}$-closed) S-lattice. Indeed, we have to prove that $\sigma_1' = \sigma_{1,S}$, where as usually, σ_1' denotes the idempotent kernel in S-mod induced by σ. Now, the extension $R \subset S$ has PDE, so we claim that $\mathbf{K}(\sigma_1') = \text{Spec}^{(1)}(S)$. Actually, if $P \in \text{Spec}^{(1)}(S)$, then $P \cap R \in \text{Spec}^{(1)}(R)$, so $p = P \cap R \in \mathbf{K}(\sigma_1)$, i.e. $P \in \mathbf{K}(\sigma_1')$. Conversely, if $P \in \mathbf{K}(\sigma_1')$, then $p = P \cap R \in \text{Spec}^{(1)}(R)$. If P does not belong to $\text{Spec}^{(1)}(S)$, then we may find $0 \subset Q \subset P$ in $\text{Spec}(S)$ and so $0 \subset Q_p \subset P_p$, all of these inclusions being strict. If we put $q = Q \cap R$, then we have either $q = 0$ or $q = p$, so $Q_p \cap R_p = 0$ or $Q_p \cap R_p = pR_p = P_p \cap R_p$. Yet, S is an R-lattice and so S_p is finitely generated over R_p, hence we have "incomparability" (INC), which yields a contradiction. This proves that divisorial S-modules are σ_1-closed and conversely. Since we know that $Cl(S) = Pic(S, \sigma_{1,S})$ and $\beta(S) = Br(S, \sigma_{1,S})$, this proves the assertion.

(9.14.) Let S be a σ-progenerator over R, e.g. let S be a σ-étale R-algebra. From (9.9.) it follows that an R-module M is σ-invertible if and only if the S-module $M \perp S$ is σ'-invertible and from this it is easy to deduce that the functor N^S_R works well on σ-invertible modules. Thus, if S is a σ-étale R-algebra, then there is a product preserving functor N^S_R : **Pic**$(S, \sigma) \to$ **Pic**(R, σ), which, by abuse of notation, defines a group morphism $N^S_R : Pic(S, \sigma) \to Pic(R, \sigma)$.

Indeed, we have $N^S_R(M \perp N) = N^S_R(M \otimes N) = N^S_R(M) \perp N^S_R(N)$ and $N^S_R(S) = R$. Note also that the composition of $N^S_R : Pic(S, \sigma) \to Pic(R, \sigma)$ with the canonical map $Pic(R, \sigma) \to Pic(S, \sigma) : [P] \to [S \perp P]$ is just multiplication by $rk_{(R, \sigma)}(S) = d$.

In particular, if $R \subset S$ are Krull domains such that S is a divisorially étale lattice over R, then we know that $Pic(R, \sigma_1) = Cl(R)$ and $Pic(S, \sigma_1) = Cl(S)$. Hence the norm yields a group morphism

$$N^S_R : Cl(S) \to Cl(R)$$

such that $Cl(R) \to Cl(S) \to Cl(R)$ is just multipliction by the (constant!) local rank of S over R.

(9.15.) This example may be generalized as follows. Call S n-<u>flat</u> over R if S_P is flat (and hence faithfully flat) over R_p for all $P \in Spec^{(n)}(S)$ and $p = P \cap R$. If S is moreover σ_n-finitely generated, then S is σ_n-flat and $(\sigma_{n,R})' = \sigma_{n,S}$, with obvious notations. Indeed, for any $p \in \mathbf{K}(\sigma_n)$ we know that S_p is finitely generated over R_p, hence (GU), (LO) and (INC) hold, so if p belongs to $Spec^{(n)}(R)$, then we may find P in $Spec^{(n)}(S)$ with $P \cap R = p$ and then $R_p \to S_p \to S_p$ is faithfully flat. It follows that the map $R_p \to S_p$ is flat, hence that S is σ_n-flat over R, indeed.

On the other hand, we know that $P \in \mathbf{K}((\sigma_{n,R})')$ if and only if $P \cap R = p \in \mathbf{K}(\sigma_{n,R})$. Moreover, if $P \in \mathbf{K}(\sigma_{n,S})$, then $\mathrm{ht}(P) \le n$, so $\mathrm{ht}(p) \le n$ as S is n-flat over R, hence P belongs to $\mathbf{K}((\sigma_{n,R})')$. Conversely, take $P \in \mathbf{K}((\sigma_{n,R})')$, so $p \in \mathbf{K}(\sigma_{n,R})$ and $\mathrm{ht}(p) \le n$. Now, $R_p \to S_p$ is flat and finitely generated and $R_p \to S_p \to S_P$ is faithfully flat. Applying (GU), (LO) and (INC) to the prime ideal P, it easily follows that $\mathrm{ht}(p) \le n$, i.e. $P \in \mathbf{K}(\sigma_{n,S})$.

(9.16.) Corollary If S is n-flat and σ_n-finitely generated, then there is an isomorphism $\mathrm{Pic}(S, \sigma_{n,R}) = \mathrm{Pic}(S, \sigma_{n,S})$.

(9.17.) Let us write $\mathrm{Cl}_n(R)$ for $\mathrm{Pic}(R, \sigma_{n,R})$; if S is n-flat and σ_n-étale over R (hence σ_n-finitely generated), then the norm map induces $N^S_R : \mathrm{Cl}_n(S) \to \mathrm{Cl}_n(R)$ and $\mathrm{Cl}_n(R) \to \mathrm{Cl}_n(S) \to \mathrm{Cl}_n(R)$ is just multiplication by the rank. Note : there is some redundancy in the above hypotheses!

The condition of being n-flat is a rather natural generalization of that of satisfying PDE and $\mathrm{Cl}_n(R)$ is a kind of "class group" adapted to this type of ring extension. As for the usual class group, there also exists a characterization of $\mathrm{Cl}_n(R)$ in terms of "divisors". Indeed, let $P(R) = K^*/R^*$, the "principal divisors" of R in K, the field of fractions of R and let $D_n(R)$ be the group of all σ_n-closed $I \subset K$ with the property that for some $J \subset K$ of the same type, we have $I * J =: Q_n(IJ) = Q_n(R)$, then from some generalities in [VV2], it follows that $\mathrm{Cl}_n(R) = D_n(R)/P(R)$. In this case (with R σ_n-closed, for simplicity's sake), we may take $J = I^{-1} = \{q \in K; qI \subset R\}$. Indeed, clearly I^{-1} is σ_n-closed, for if $Lq \subset I^{-1}$ for some $q \in K$ and $L \in \mathbf{L}(\sigma_n)$, then $LqI \subset R$ so $qI \subset R$ as R is σ_n-closed and $q \in I^{-1}$, proving the assertion. Next, from $Q_n(IJ) = R$, we get that for each $p \in \mathbf{K}(\sigma_n)$ that $I_p J_p = R_p$. Now $IJ \subset Q_n(IJ) = R$, so $J \subset I^{-1}$ and so $J_p \subset (I^{-1})_p$. We thus obtain $R_p = I_p J_p \subset I_p(I^{-1})_p = (II^{-1})_p \subset R_p$, i.e. $(II^{-1})_p = R_p$, hence $Q_n(II^{-1}) = R$.

(9.18.) One may derive similar results for relative Brauer groups. Here one proves that if S is σ-étale over R, then the norm induces a functor

$$N^S_R : \underline{Az}(S, \sigma) \to \underline{Az}(R, \sigma),$$

where $\underline{Az}(R, \sigma)$ denotes the category of σ-Azumaya algebras over R. Here one uses the fact, mentioned in (9.9.), that σ-Azumaya algebras behave well with respect to descent. The only detail we have to check is whether σ'-progenerators resp. endomorphism rings of σ'-progenerators are mapped onto σ-progenerators resp. endomorphism rings of σ-progenerators by N^S_R.

This may be checked as follows. We may obviously assume that S is of the form R x ... x R (d times), applying some straightforward descent argument. Now, if E is a σ'-progenerator over S, then $E = E_1 \times ... \times E_d$, where each E_i is a σ-finitely generated R-module, hence $E_1 \otimes ... \otimes E_d$ and $N^S_R(E)$ are σ-finitely generated over R. The definition of of N^S_R yields that $N^S_R(E)_p = N^{S'}_{R'}(E_p)$ for each $p \in \mathbf{K}(\sigma)$, where $R' = R_p$ resp. $S' = S_p$ as usually, so $N^S_R(E)$ is a σ-progenerator over R, indeed. We leave it as an exercise in descent theory (or refer to [KO2, Prop.4.4.] for a similar proof in the absolute case) to see that $N^S_R(End_S(E)) = End_R(N^S_R(E))$ for a σ-progenerator E over S.

Hence:

(9.19.) Proposition If S is σ-étale over R, then N^S_R induces a group morphism

$$N^S_R : Br(S, \sigma) \to Br(R, \sigma).$$

It follows that any $[A] \in Br(R, \sigma)$, which may be split by a σ-étale S of local rank d over R has exponent d in $Br(R, \sigma)$.

Note also that for a pair of Krull domains $R \subset S$ such that S is a divisorially étale R-lattice, there exists a group morphism

$$N^S_R : \beta(S) \to \beta(R).$$

(9.20.) Let us conclude by briefly indicating what happens in the geometric situation. Let $R \to S$ be an injective morphism of noetherian domains, then for all ideals I of R this yields a group morphism $\mathrm{Pic}(R, \sigma_I) \to \mathrm{Pic}(S, (\sigma_I)')$. Let $J = SI$, the S-ideal generated by I, then we claim that $(\sigma_I)' = \sigma_J$. Indeed, $P \in \mathbf{K}((\sigma_I)')$ if and only if $(\sigma_I)'(S/P) = 0$, i.e. $I \subset P$. This proves that $P \in \mathbf{K}(\sigma_J)$ and conversely. Of course $X_R(I) = \mathbf{K}(\sigma_I)$ and $X_S(J) = \mathbf{K}(\sigma_J)$ and conversely. Of course $X_R(I) = \mathbf{K}(\sigma_I)$ and $X_S(J) = \mathbf{K}(\sigma_J)$. Now, the above morphism yields a map $f : \mathrm{Spec}(S) \to \mathrm{Spec}(R)$ (which is continuous, of course) and one easily verifies that $f^{-1}(X_R(I)) = X_S(J)$, so the above map reduces to the usual morphism $\mathrm{Pic}(X_R(I)) \to \mathrm{Pic}(X_S(J))$, using the isomorphisms $\mathrm{Pic}(R, \sigma_I) \cong \mathrm{Pic}(X_R(I))$ and $\mathrm{Pic}(S, \sigma_J) \cong \mathrm{Pic}(X_S(J))$. Similarly, for Brauer groups we get a map $\mathrm{Br}(X_R(I)) \to \mathrm{Br}(X_S(J))$, using the analogous isomorphisms $\mathrm{Br}(R, \sigma_I) \cong \mathrm{Br}(X_R(I))$ resp. $\mathrm{Br}(S, \sigma_J) \cong \mathrm{Br}(X_S(J))$. It is easy to check that the induced map $f|X_S(J) : X_S(J) \to X_R(I)$ is étale of finite type if and only if S is σ_I-étale over R (hence also σ_I-finitely generated by definition). It then follows that we then have a norm map

$$N^S_R : \mathrm{Pic}(X_S(J)) = \mathrm{Pic}(S, \sigma_I) \to \mathrm{Pic}(R, \sigma_I) = \mathrm{Pic}(X_R(I)),$$

which coincides with the usual corestriction on Picard groups.
The cohomological fact that $\mathrm{cores} \circ \mathrm{res} : \mathrm{Pic}(X_R(I)) \to \mathrm{Pic}(X_S(J)) \to \mathrm{Pic}(X_R(I))$ is just multiplication by the the degree then follows directly from the foregoing. The analogous map $N^S_R : \mathrm{Br}(X_S(J)) \to \mathrm{Br}(X_R(I))$ may be treated similarly.

(9.21.) Let us now introduce Hecke actions and show how these relate to the previously introduced norm maps.
Throughout X will denote a separated integral scheme, which is locally of

finite type with respect to some generically closed subset Y of X. As before, we let $\underline{O}_Y = \underline{O}_X|Y$.

If $\underline{A}$ is a commutative quasicoherent sheaf of $\underline{O}_X$-algebras on X, then we say that a sheaf of $\underline{A}$-modules E on X is Y-<u>closed</u> if it is quasicoherent over $\underline{A}$ (and hence over $\underline{O}_X$!) and if the canonical map $E \to Q_Y(E)$ (which is a morphism of sheaves of $\underline{A}$-modules) is an isomorphism. We say that E is Y-<u>invertible</u> if we may find an open covering $\{U_i; i \in I\}$ of Y in X such that $E|U_i \cap Y \cong \underline{A}|U_i \cap Y$ for all $i \in I$. Of course, all of these U_i may be chosen to be affine. It follows in particular that E is Y-invertible if and only if $E|Y$ is a locally free sheaf of $\underline{O}_Y$-modules of rank 1 on Y.

The tensor product induces a group operation on the set of isomorphism classes of Y-closed Y-invertible sheaves of $\underline{O}_X$-modules and we denote the group thus obtained by Pic(X, $\underline{A}$; Y). Using (6.17.) and (6.21.) it is fairly easy to show that Pic(X, $\underline{A}$; Y) = H^1(Y, $(\underline{A}|Y)^*$) = Pic(Y, $\underline{A}|Y$), so in particular, if $\underline{A} = \underline{O}_X$, then by (7.10.), the group Pic(X, $\underline{O}_X$; Y) coincides with the relative Picard group of (X, $\underline{O}_X$) with respect to Y, also denoted by Pic(X, Y).

The corresponding local notion is just $\mathrm{Pic}_R(A, \sigma) =: \mathrm{Pic}(\mathrm{Spec}(R), \underline{O}_A; \mathbf{K}(\sigma))$ for any idempotent kernel functor of finite type σ in R-mod and any commutative algebra A over R.

(9.22.) Proposition Let (X, Y) be as before and let $\underline{A}$ be a quasicoherent sheaf of $\underline{O}_X$-algebras, then a (Y-closed) sheaf of $\underline{A}$-modules E on X is Y-invertible if and only if the following conditions are satisfied:

(9.22.1.) there is an isomorphism $E_x \cong \underline{A}_x$ for all $x \in X$;

(9.22.2.) $\Gamma(U, E)$ is a $\sigma_{U \cap Y}$-finitely generated $\Gamma(U, \underline{A})$-module for all U in some covering of Y by open affines of X.

Proof If E is a Y-closed Y-invertible sheaf of A-modules, then on any U in some open affine covering of Y in X, there is an isomorphism $E|U \cap Y \cong$

$\underline{A}|U \cap Y$. Now, on $U = \text{Spec}(R)$, clearly E is of the form $E|U = \underline{Q}_M$, where M is some module over $A = \Gamma(U, \underline{A})$. It follows that $Q_\sigma(M) = Q_\sigma(A)$, where $\sigma = \sigma_{U \cap Y}$ in R-mod. Viewing this isomorphism as an identification and using the fact that σ has finite type yields that M is σ-finitely generated over A. The assertion (1) is then of course trivially satisfied, proving that the two statements (1), (2) are necessary.

Conversely, it is clear that we may work locally over open affines $U = \text{Spec}(R)$, with $Y \cap U = \mathbf{K}(\sigma)$. Assume that M is a σ-finitely generated A-module, where we put $A = \Gamma(U, \underline{A})$, and that $M_p \cong A_p$ for all $p \in \mathbf{K}(\sigma)$. Assume $\psi : A_p \to M_p$ is an isomorphism induced from A, i.e. $\psi = \phi_p$ for some $\phi : A \to M$ given by $\phi(a) = am$ for all $a \in A$ and some fixed $m \in M$. Since M is σ-finitely generated, we may find a finitely generated A-submodule $N \subset M$ with the property that M/N is σ-torsion. Pick a set of generators S of N, then the localized set S_q generates $M_q = N_q$ for all q in $\mathbf{K}(\sigma)$ such that $S_q \subset \phi(F)_q$ for some finitely generated R-submodule F of A, i.e. such that $(Rs/Rs \cap \phi(F))_q = 0$ for all $s \in S$. If we fix F, then the set of all q in $\text{Spec}(R)$ with the property that $(Rs/Rs \cap \phi(F))_q = 0$ is open in $\text{Spec}(R)$, so it follows that $\phi_q : A_q \to M_q$ is a surjective map for all $q \in \mathbf{K}(\sigma) \cap V$, where V is an affine open subset of $\text{Spec}(R)$. It follows that $\underline{Q}_A|\mathbf{K}(\sigma) \cap V \cong \underline{Q}_M|\mathbf{K}(\sigma) \cap V$ and from this it follows that E is Y-invertible, indeed.

(9.23.) Let G be an arbitrary finite group and for any subgroup H of G, denote by $\mathbf{Z}[G/H]$ the free $\mathbf{Z}$-module with basis G/H. We let $\mathbf{H}_G$ denote the <u>Hecke category</u> over G, i.e. with objects $\mathbf{Z}[G/H]$ and morphisms the $\mathbf{Z}[G]$-module (or "G-module") homomorphisms. One may prove that

$$\text{Hom}_G(\mathbf{Z}[G/H], \mathbf{Z}[G/K]) = \mathbf{Z}[H\backslash G/K],$$

for any pair of subgroups H, K of G.

Indeed, for any $x \in G$, define $\Phi(HxK) \in \text{Hom}_G(\mathbf{Z}[G/H], \mathbf{Z}[G/K])$ by

$$\Phi(HxK)(gH) = \sum guxK = \sum gu'K,$$

for any $g \in G$, where u runs through a collection of coset representatives of $H/H \cap xKx^{-1}$ and u' through representatives of HxK/K. We leave it as an easy exercise to the reader to prove that F actually defines a bijection between $\mathbf{Z}[H\backslash G/K]$ and $\mathrm{Hom}_G(\mathbf{Z}[G/H], \mathbf{Z}[G/K])$.

We will also have to use the extended Hecke category $(\mathbf{H}_G)^\wedge$, whose objects are finite direct sums of objects in $\mathbf{H}_G$ and where morphisms are defined in the obvious way.

Let H, K, L be subgroups of G, then for any morphism $f : \mathbf{Z}[G/H] \to \mathbf{Z}[G/K]$ and any morphism $g : \mathbf{Z}[G/K] \to \mathbf{Z}[G/L]$ we obtain a composition gf, which induces a bilinear map

$$\mathbf{Z}[H\backslash G/K] \times \mathbf{Z}[K\backslash G/L] \; \to \; \mathbf{Z}[H\backslash G/L],$$

which sends (x, y) to $xy = \Phi^{-1}(\Phi(y)\Phi(x))$.

We then have

(9.24.) Lemma For any $x, y \in G$ and any subgroups H, K, L of G, we have

$$(HxK)(KyL) \; = \; \sum n(x, y; z)HzL,$$

where z runs through a set of double coset representatives of $H\backslash G/L$ and where $n(x, y; z) = |(HxK \cap zLy^{-1}K)/K|$.

Proof Φ maps (HxK)(KyL) to a map sending H to $\sum uvL = \sum n(x, y; z)zL$, where u runs through a set of representatives of HxK/K, v through a set of representatives of KyL/L and z through a set of double coset representatives of $H\backslash G/L$. It follows that

$$n(x, y; z) \; = \; |\{(uK, vL); uK \in HxK/K, vL \in KyL, uvL = zL\}|$$

$$= \Sigma_u|\{vL; vL \in KyL/L, uvL = zL\}|$$
$$= \Sigma_u|(u^{-1}zL \cap KyL)/L|,$$

where u runs through a set of representatives of HxK/K.

But $|(u^{-1}zL \cap KyL)/L| = 1$ if $u \in zLy^{-1}K$ and 0 otherwise, hence we find that

$n(x, y; z) = |\{uK; uK \in HxK/K, u \in zLy^{-1}K\}| = |(HxK \cap zLy^{-1}K)/K|$.

(9.25.) Proposition Under the above assumptions, suppose there is a group G acting as $\underline{O}_X$-automorphisms on $\underline{A}$, then there exists a contravariant additive functor

$$T : \mathbf{H}_G \rightarrow \textbf{(abelian groups)},$$

which sends $\mathbf{Z}[G/H]$ to $H^1(Y, ((\underline{A}|Y)^H)^*) = \mathrm{Pic}(Y, (\underline{A}|Y)^H)$.

Proof (cf. [RS]) Of course $\underline{A}^H$ is defined by putting $\Gamma(U, \underline{A}^H) = \Gamma(U, \underline{A})^H$, etc. for any $U \subset X$. Now, if M is denotes an arbitrary G-module, then clearly there is a contravariant additive functor $\mathbf{H}_G \rightarrow$ **(abelian groups)**, which sends $\mathbf{Z}[G/H]$ to $M^H = \mathrm{Hom}_H(\mathbf{Z}, M) = \mathrm{Hom}_G(\mathbf{Z}G \otimes_{\mathbf{Z}H} \mathbf{Z}, M)$.

But $(\mathbf{Z}[G/H], M) \rightarrow M^H$ even defines a bifunctor, so there is a contravariaant additive functor (for $\underline{any}\ \underline{A}$ and X!)

$$\mathbf{H}_G \rightarrow \underline{\mathbf{S}}(X) : \mathbf{Z}[G/H] \rightarrow \underline{A}^H.$$

Restricting to Y and noting that $((\underline{A}|Y)^H)^* = ((\underline{A}|Y)^*)^H$, it easily follows that T is well defined.

(9.26.) Notes.

(9.26.1.) It is clear that T extends to a contravariant functor

$$T^\wedge : (\mathbf{H}_G)^\wedge \rightarrow \textbf{(abelian groups)}.$$

(9.26.2.) Denoting T by T_Y if we want to stress its dependency upon Y, then taking restrictions yields for any pair $Y' \subset Y$ a natural transformation of functors $\eta_{Y, Y'} : T_Y \to T_{Y'}$, i.e. these functors are compatible with the natural maps

$$H^1(Y, ((\underline{A}|Y)^H)^*) \to H^1(Y', ((\underline{A}|Y')^H)^*).$$

(9.27.) Proposition Let $f : Z \to X$ be an affine morphism of schemes and let Y be a generically closed subset of X, such that X is locally of finite type with respect to Y. Assume that $(f_*\underline{O}_Z)_x$ is semilocal for all $x \in Y$, then for any sheaf of $\underline{O}_Z$-modules E the following assertions are equivalent:

(9.27.1.) $[f_*E] \in Pic(X, f_*\underline{O}_Z; Y)$;

(9.27.2.) $[E] \in Pic(Z, \underline{O}_Z; f^{-1}(Y)) = Pic(Z, f^{-1}(Y))$.

Proof It is fairly trivial to see that $f^{-1}(Y)$ is generically closed and that (1) $\Rightarrow$ (2).

To prove that (2) $\Rightarrow$ (1), note that the problem is local, so we may assume that f is of the form $Spec(S) \to Spec(R)$ associated to some ring homomorphism $R \to S$ and that $Y \subset X$ is of the form $\mathbf{K}(\sigma)$ for some idempotent kernel functor of finite type in R-mod. It is easy to see that in this case $f^{-1}(Y) = \mathbf{K}(\sigma')$, the generically closed subset of $Spec(S)$, associated to the idempotent kernel functor σ'. Of course σ' also has finite type (in S-mod!). Any $f^{-1}(Y)$-invertible sheaf E on Z is then of the form $\underline{O}_N$ for some σ'-generated S-module N. Note also that $f_*\underline{O}_N = \underline{O}_{N'}$, where N' is N viewed as an R-module and $f_*\underline{O}_Z = \underline{O}_{S'}$ (idem!). Now, $(f_*\underline{O}_N)_p$ is a finitely generated $(f_*\underline{O}_Z)_p$-module for any $p \in Y = \mathbf{K}(\sigma)$. Since E is $f^{-1}(Y)$-invertible and $(f_*\underline{O}_Z)_p$ is semilocal by assumption, it follows that $(f_*\underline{O}_N)_p = (f_*\underline{O}_Z)_p$ up to isomorphism. Finally, since N is σ'-finitely generated over S, it follows that $N' = \Gamma(X, f_*E)$ is σ-finitely generated over $S = \Gamma(X, f_*\underline{O}_Z)$. The conclusion now follows immediately from (9.22.).

(9.28.) Corollary The above assumptions also imply that

$$\text{Pic}(X, f_*\underline{O}_Z; Y) \;=\; \text{Pic}(Z, \underline{O}_Z; f^{-1}(Y)).$$

We are now ready to prove one of the main results of this Section.

(9.29.) Theorem Let A be a (commutative) R-algebra and let G be a finite group of R-automorphisms of A. Suppose that σ is an idempotent kernel functor of finite type in R-mod, then there is a contravaraiant additive functor

$$T \;:\; \mathbf{H_G} \;\rightarrow\; \textbf{(abelian groups)}$$

which sends $\mathbf{Z}[G/H]$ to $\text{Pic}(A^H, \sigma)$.

Proof The proof of this result is a modification of that of [RS, Theorem 3.4.] which it generalizes.

For any subgroup H of G the inclusions $A^G \subset A^H \subset A$ induce a diagram of scheme morphisms

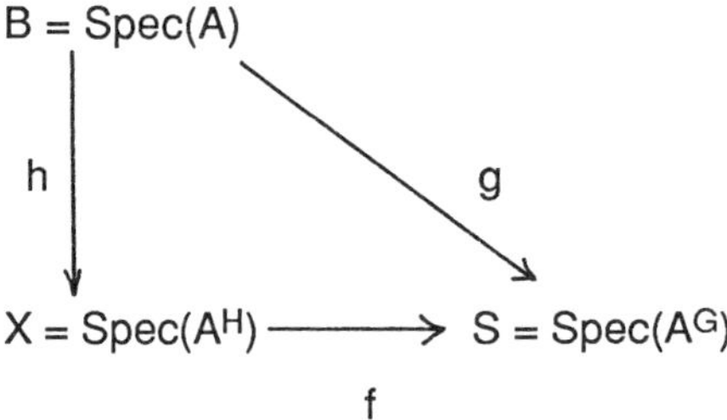

Let $\underline{O}_A$ be the structure sheaf on B and $\underline{O}_{A^H}$ that on X, then we claim that $(g_*\underline{O}_A)^H = f_*\underline{O}_{A^H}$. Indeed, for any $s \in A^G$, let us denote by $S(s) \subset S$ resp. $B(s) \subset B$, $X(s) \subset X$ the corresponding basic open subset, then we have

$\Gamma(S(s), (g_*\underline{O}_A)^H) = \Gamma(S(s), g_*\underline{O}_A)^H = \Gamma(g^{-1}(S(s)), \underline{O}_A)^H = \Gamma(B(s), \underline{O}_A)^H = (A_s)^H,$

resp.

$\Gamma(S(s), f_*\underline{O}_AH) = \Gamma(f^{-1}(S(s)), \underline{O}_A{}^H) = \Gamma(X(s), \underline{O}_A{}^H) = (A^H)_s.$

We leave it as an easy verification to the reader to check that $(A_s)^H = (A^H)_s$ (since G and hence H is finite!). This proves the result.

Let $\underline{A} = g_*\underline{O}_A$, then we just found that $\underline{A}^H = f_*\underline{O}_AH$ for any subgroup H of G. We will prove in a moment that $(f_*\underline{O}_AH)_p$ is semilocal for all $p \in S$, so we may apply (9.28.) to derive that

$$Pic(S, f_*\underline{O}_AH; \mathbf{K}(A^G, \sigma)) = Pic(X, \underline{O}_AH; \mathbf{K}(A^H, \sigma)),$$

where $\mathbf{K}(A^H, \sigma) = f^{-1}(\mathbf{K}(A^G, \sigma))$. But the latter group is just $Pic(A^H, \sigma)$, so we may apply (9.25.) to derive a functor $T : \mathbf{H}_G \to$ **(abelian groups)**, which sends $\mathbf{Z}[G/H]$ to $Pic(S, \underline{A}^H; \mathbf{K}(A^G, \sigma)) = Pic(A^H, \sigma)$, indeed. It thus remains to verify that $(f_*\underline{O}_AH)_p$ is semilocal for all $p \in Spec(A^G)$.

First note that A is integral over A^G, since every $a \in A$ is a zero of the monic polynomial $p_a(t) = \prod_{g \in G}(t - g(a))$ with coefficients in A^G, hence a fortiori A^H is integral over A^G.

Now, if $p \in S = Spec(A^G)$, then every maximal ideal of $(A^H)_p$ contains the maximal ideal of $(A^G)_p$ (by "going up"), so it suffices to prove that $f^{-1}(p)$ is finite. However, since h is surjective (by "lying over" applied to the integral extension $A^H \subset A$) it is enough to show that $P = g^{-1}(p)$ is finite.

But, G acts on P in the obvious way and we claim that this action is transitive (proving the assertion, since G is finite!). Indeed, let p_1 and p_2 be different primes in P, and choose $x \in p_1$, then $\prod_{g \in G}g(x) \in p_1 \cap A^G = p \subset p_2$, so $g^{-1}(x) \in p_2$ for some $g_2 \in G$. It thus follows that $p_1 \subset \cup_{g \in G}g(p_2)$, so

$p_1 \subset g(p_2)$ for some g, by [AM, (1.11.)]. But then $p_1 = g(p_2)$, by "incomparability", since $g(p_2) \cap A^G = p = p_1 \cap A^G$, proving the assertion.

(9.30.) Note The conditions required to make (9.27.) and (9.28.) work are also satisfied in other situations, for example if $f : Z \to X$ is of Y-finite type, i.e. if f is affine and if for all open subsets $\mathrm{Spec}(R) \subset X$ in some open affine covering of X with $\mathrm{Spec}(S) = f^{-1}(\mathrm{Spec}(R)) \subset Z$, we have that the ring morphism $R \to S$ induced by $f|\mathrm{Spec}(S)$ makes S into a $\sigma_{\mathrm{Spec}(R) \cap Y}$-finitely generated R-module. In this case for any $x \in Y$, every maximal ideal of $(f_* \underline{O}_Z)_x$ contains the image of the maximal ideal of $\underline{O}_{X,x}$ by "going up", hence these ideals correspond bijectively to the maximal ideals in a finite dimensional algebra, i.e. $(f_* \underline{O}_Z)_x$ is semilocal, indeed.

For other sufficient conditions, we refer to [Ve6].

Note also that in (9.29.) usually R will be σ-noetherian and A a σ-finitely generated R-algebra. In this case A^H will be σ-finitely generated too, for all $H < G$, which allows us to slightly simplify the proof given above in this case.

(9.31.) There are some special morphisms in the category $\mathbf{H}_G$, playing an important role in its structure theory. These are:

(9.31.1.) $T^K = T^K_H = H1K$ for $H < K < G$;

(9.31.2.) $R_H = R_H{}^K = K1H$ for $H < K < G$;

(9.31.3.) $S^g = S^g_H = HgH^g$ for $H < G$, $g \in G$ and $H^g = g^{-1}Hg$;

(9.31.4.) $I = I_H = H1H$ for $H < G$.

With these notations, we have the following formula, which is essentially due to Mackey and which explains the importance of a careful study of the previous morphisms: if H, K are subgroups of G, $g \in G$ and if we put $D = H^g \cap K$, then

$$HgK = (HgH^g)(H^g1D)(D1K) = S^g{}_H R_D T^K.$$

It follows that almost all calculations and results on Hecke categories may be reduced to a study of the elementary operators T^K, R_H and S^g for $H, K < G$ and g in G.

(9.32.) Let A be a commutative R-algebra, G a finite group of R-automorphisms of A and let σ an idempotent kernel functor of finite type in R-mod. Applying the functor T to the morphism $T^K{}_H$ and $R_H{}^K$ for $H < K < G$ yields group homomorphisms

$$res^K{}_H = T(T^K{}_H) \ : \ Pic(A^K, \sigma) \ \to \ Pic(A^H, \sigma)$$

resp.

$$cores_H{}^K = T(R_H{}^K) \ : \ Pic(A^H, \sigma) \ \to \ Pic(A^K, \sigma).$$

The morphism I_H ($H < G$) is mapped by T to the identity map on $Pic(A^H, \sigma)$. Using (9.24.) it is easy to verify that $R_H{}^K T^K{}_H = |K/H| I_K$. Indeed, we have $(K1H)(H1K) = \sum n(1, 1; z) KzK$, where z runs through a set of representatives of $K \backslash G / K$ and $n(1, 1; z) = |(K1H \cap zK1H)/H| = |(K \cap zK)/H|$, hence $n(1, 1; z) = |K/H|$ or 0 depending on whether $z \in K$ or not.
Applying T then yields that the composition

$$cores_H{}^K res^K{}_H \ : \ Pic(A^K, \sigma) \ \to \ Pic(A^H, \sigma) \ \to \ Pic(A^K, \sigma)$$

is just multiplication by $|K/H|$ (note that $R_H{}^K T^K{}_H = T^K{}_H \circ R_H{}^K$, omitting Φ!). In particular, if $H = \{e\}$, we write $res^K = res^K{}_H$ resp. $cores^K = cores_H{}^K$, hence $cores^K res^K = |K|$!

(9.33.) We claim that for $H < K < G$ the morphism $\mathrm{res}^K{}_H : \mathrm{Pic}(A^K, \sigma) \to \mathrm{Pic}(A^H, \sigma)$ is just the map $\mathrm{Pic}(A^K, \sigma) \to \mathrm{Pic}(A^H, \sigma)$ induced by the inclusion $A^K \subset A^H$. Indeed, without loss of generality, we may assume $H = \{e\}$ and $K = G$, i.e. we consider the map $\mathrm{res} = \mathrm{res}^G : \mathrm{Pic}(C, \sigma) \to \mathrm{Pic}(A, \sigma)$ where $C = A^G$.

Let $g : B = \mathrm{Spec}(A) \to S = \mathrm{Spec}(C)$ be induced by $A^G = C \subset A$ as in (9.29.), then by (9.28.) sending $[E]$ to $[g_*E]$ defines an isomorphism

$$w \; : \; \mathrm{Pic}(B, \underline{O}_A; \mathbf{K}(A, \sigma)) \; \to \; \mathrm{Pic}(S, \underline{A}; Y),$$

where $\underline{A} = g_*\underline{O}_A$ and $Y = \mathbf{K}(C, \sigma)$. Since $\underline{A}$ is quasicoherent by the proof of (9.29.), there also is an isomorphism

$$u \; : \; \mathrm{Pic}(S, \underline{A}; Y) \; \to \; H^1(Y, (\underline{A}|Y)^*) \; = \; \mathrm{Pic}(Y, \underline{A}|Y)$$

sending $[E] \in \mathrm{Pic}(S, \underline{A}; Y)$ to the class of $E|Y$ in $H^1(Y, (\underline{A}|Y)^*)$ or $\mathrm{Pic}(Y, \underline{A}|Y)$. Finally, since $\underline{A}^G = \underline{O}_C$ is also quasicoherent, there is an isomorphism

$$v \; : \; \mathrm{Pic}(Y, (\underline{A}|Y)^G) \; = \; H^1(Y, (\underline{A}^G|Y)^*) \; \to \; \mathrm{Pic}(S, \underline{A}^G; Y),$$

where we used the fact that $\underline{A}^G|Y = (\underline{A}|Y)^G$, essentially because for any σ we have $Q_\sigma(A^G) = Q_\sigma(A)^G$ on the algebra level. This yields a commutative diagram

$$\mathrm{Pic}(Y, (\underline{A}|Y)^G) \to \mathrm{Pic}(S, \underline{A}^G; Y) \; = \; \mathrm{Pic}(S, \underline{O}_C; Y) \; = \; \mathrm{Pic}(C, \sigma)$$

$$\tau \downarrow \qquad\qquad\qquad\qquad\qquad\qquad\qquad\qquad\qquad\qquad \downarrow \mathrm{res}$$

$$\mathrm{Pic}(Y, \underline{A}|Y) \leftarrow \mathrm{Pic}(S, \underline{A}; Y) \leftarrow \mathrm{Pic}(B, \underline{O}_A; \mathbf{K}(A, \sigma)) = \mathrm{Pic}(A, \sigma),$$

where the morphisms are the indicated ones or either trivial, except for the morphism τ : Pic(Y, $(\underline{A}|Y)^G$) $\to$ Pic(Y, $\underline{A}|Y$), which is given by (9.25.). But τ is given by [E] $\to$ [E $\otimes_Y (\underline{A}|Y)$], so one easily checks the map [M] $\to$ [A $\perp$ M] from Pic(C, σ) to Pic(A, σ) (modified tensor product over R!) also to make the diagram commute, i.e. it coincides with res, as claimed.

(9.34.) Argueing similarly for cores : Pic(A, σ) $\to$ Pic(C, σ), we get a commutative diagram

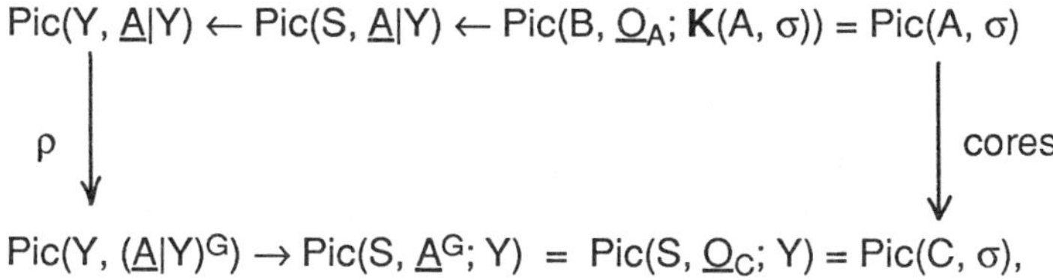

where ρ : Pic(Y, $\underline{A}|Y$) $\to$ Pic(Y, $(\underline{A}|Y)^G$) is determined by (9.25.) again. Now assume that S is a σ-Galois extension of R and let A = S, hence C = $A^G = S^G$ = R. Since in this case $\rho([E]) = [(\otimes_{g \in G} E_g)^G]$, where E_g is the g-twisted version of E and G acts in the obvious way, it follows that cores and N^S_R coincide in this case, due to (9.12.).

(9.35.) As an application, with notations as before, let p be a positive prime in **Z** which does not divide |K/H| for some H < K < G, then we claim that the canonical map Pic(A^K, σ)$_{(p)}$ $\to$ Pic(A^H, σ)$_{(p)}$ is injective, where the p-component (-)$_{(p)}$ is defined to consist of all x with p|ord(x) or x = 0. Indeed, we have seen in (9.33.) that the canonical map Pic(A^K, σ) $\to$ Pic(A^H, σ) coincides with res^K_H. But $\text{cores}_H{}^K \circ \text{res}^K_H$ is just multiplication by |K/H|, hence res^K_H is injective when restricted to the p-component of Pic(A^K, σ) indeed, since p does not divide |K/H| by assumption. In particular, if S is a σ-Galois extension of R with Galois group G and p does not divide the order |G| of G, then

$$\text{Pic}(R, \sigma)_{(p)} \subset \text{Pic}(S, \sigma)_{(p)}.$$

(9.36.) Let us now assume $\{H(i); 1 \leq i \leq n\}$ to be a family of subgroups of $H < G$ with $\text{g.c.d.}(\{|H/H(i)|; 1 \leq i \leq n\} = 1$. Choose integers $r_i \in \mathbf{Z}$ such that $\sum r_i|H/H(i)| = 1$, then there are morphisms

$$i = \oplus_i T^H{}_{H(i)} \; : \; \oplus_i \mathbf{Z}[G/H(i)] \to \mathbf{Z}[G/H]$$

resp.

$$j = (r_i R_{H(i)}{}^H)_i \; : \; \mathbf{Z}[G/H] \to \oplus_i \mathbf{Z}[G/H(i)]$$

in $(\mathbf{H}_G)^{\wedge}$. An easy calculation shows that $i \circ j = 1$, by the choice of the r_i. Applying $T^{\wedge}$ yields that $T(j) \circ T(i) = \text{id}_{\text{Pic}(A^H, \sigma)}$, hence $T(i)$ is injective. But $T(i)$ is just the obvious map

$$\text{Pic}(A^H, \sigma) \subset \prod \text{Pic}(A^{H(i)}, \sigma).$$

In particular, if S is σ-Galois over R with group G and $\{H(i); 1 \leq i \leq n\}$ a family of subgroups of G with $\text{g.c.d.}(\{|G/H(i)|; 1 \leq i \leq n\}) = 1$, then we get an injective map

$$\text{Pic}(R, \sigma) \subset \prod \text{Pic}(S^{H(i)}, \sigma).$$

(9.37.) Let us now consider the case of Krull domains as a concrete example. Let R be a Krull domain, then we denote by σ_1 or $\sigma_{1,R}$ the idempotent kernel functor in R-mod associated to $\text{Spec}^{(1)}(R)$. We know that $\text{Pic}(R, \sigma_1) = \text{Cl}(R)$. If S is another Krull domain which is a (divisorial) R-lattice, then we have seen in (9.13.) that $\sigma_{1,S} = (\sigma_{1,R})'$, so $\text{Pic}(S, \sigma_1) = \text{Cl}(S)$.

Assume that G is a (finite) group of R-automorphisms of S, then S^H is also a Krull domain, as $S^H = S \cap L^H$, where H acts in the obvious way on L, the field of fractions of S. Moreover, S^H is σ_1-finitely generated over R (hence it is a divisorial R-lattice by (8.47.)), being a submodule of S, which is $\sigma_{1,R}$-finitely generated, hence $\sigma_{1,R}$-noetherian. We thus find that $\text{Pic}(S^H, \sigma_{1,R}) = Cl(S^H)$. Hence:

(9.38.) Proposition Let $R \subset S$ be an extension of Krull domains making S into a (divisorial) R-lattice, then for any finite group of R-automorphisms G of S, there is a contravariant additive functor

$$T \: : \: H_G \: \to \: \textbf{(abelian groups)}$$

sending $\mathbb{Z}[G/H]$ to $Cl(S^H)$.

(9.39.) Corollary Let H be a subgroup of G and p a positive prime in $\mathbb{Z}$ such that p does not divide $|G/H|$, then the map $Cl(S^G)_{(p)} \to Cl(S^H)_{(p)}$ is injective. In particular, if S is divisorially Galois over R with Galois group G, then the map $Cl(R)_{(p)} \to Cl(S)_{(p)}$ is injective for all p not dividing $|G|$.

We leave it to the reader to formulate (9.36.) in this special case.

(9.40.) Let us now briefly sketch how the foregoing relates to Yoshida's theory of G-functors developed in [Yo]. The next paragraphs will not be needed in the sequel.

Let G be a finite group for a moment, then for any commutative ring k a G-functor over k if a quadruple $(\textbf{a}, \tau, \rho, \sigma)$, where $\textbf{a}$ is a map from sub(G), the set of subgroups of G to k-mod, and where τ, ρ, σ are connectors, i.e. for all $H < K$ in sub(G) and all $g \in G$, there are given k-linear maps $\tau^K_H \: : \: \textbf{a}(H) \to \textbf{a}(K) \: : \: a \to a^K$ resp. $\rho_H^K \: : \: \textbf{a}(K) \to \textbf{a}(H) \: : \: a \to a_H$ and $\sigma^g_H \: : \: \textbf{a}(H) \to$

$\mathbf{a}(H^g) : a \to a^g$, satisfying certain compatibilities described in detail in [Yo] (e.g. $\tau^H{}_H = id_H$, $\tau^L{}_K\tau^K{}_H = \tau^L{}_H$, etc.) and the so-called <u>Mackey axiom</u>. This axiom states that for all H, K < L in sub(G) and all a $\in$ H we have

$$(a^L)_K = \sum (a^g{}_{H^g \cap K})^K,$$

where g runs through a set of double coset representatives for H\L/K. We leave it to the reader to formulate this axiom directly in terms of the connectors.

(9.41.) As an example, let V be a kG-module and let g $\in$ G resp. H < K in sub(G). Put $\mathbf{a}(H) = V^H$, then we may define a G-functor over k by putting

(9.41.1.) $\tau^K{}_H : \mathbf{a}(H) \to \mathbf{a}(K) : v \to \sum g(v)$ (g runs through H\K);

(9.41.2.) $\rho_H{}^K : \mathbf{a}(K) \to \mathbf{a}(H) : v \to v$;

(9.41.3.) $\sigma^g{}_H : \mathbf{a}(H) \to \mathbf{a}(Hg) : v \to v(g)$.

The maps τ, ρ and σ are called <u>corestriction</u>, <u>restriction</u> resp. <u>conjugation</u>. One easily checks these to be well-defined and to satisfy the axioms for a G-functor. However, one may verify that $(\mathbf{a}, \tau, \rho, \sigma)$ satisfies in this case also the so-called cohomological axiom (C) :

(C) for any H < K in sub(G) and any b $\in$ $\mathbf{a}(K)$ we have $(b_H)^K = |K/H|b$.

In general, a G-functor satisfying axiom (C) is called a <u>cohomological</u> G-functor.

The collection of G-functors resp. of cohomological G-functors may be made into a category $\mathbf{M}_k(G)$ resp. $\mathbf{M}_k(G)^c$ in a straightforward way, we refer to [Yo] for details.

(9.42.) Let us return to the Hecke category $\mathbf{H}_G$ of G. We may generalize the foregoing construction to an arbitrary ring k (i.e. objects are the k[G/H], where H belongs to sub(G), etc.) and we then obtain the Hecke category $\mathbf{H}_{kG}$ of G over k. The special objects $T^K{}_H$, $R_H{}^K$, $S^g{}_H$ and I_H may still be defined together with the relation

$$HgK \;=\; S^g{}_H R_D T^K.$$

From this one easily deduces the relation

$$T^L{}_H R^L{}_K \;=\; \Sigma_{H\backslash L/K} S^g{}_H R_{Hg \,\cap\, K} T^K,$$

where H, K < L and where g runs through a set of double coset representatives for H\L/K. It is also straightforward to prove that $R_H{}^K T^K{}_H = |K/H| I_K$ for H < K in Sub(G), as well as some compatibility relations such as $T^H{}_H = I_H$, $T^K{}_H T^L{}_K = T^L{}_H$ for H < K < L, ... The fact that these relations have the same form as those mentioned for G-functors provides evidence that there should be some connection between Hecke categories and G-functors. More precisely:

(9.43.) Theorem (T. Yoshida [Yo]) For any cohomological G-functor $(\mathbf{a}, \tau, \rho, \sigma)$ there exists a k-additive functor $A : \mathbf{H}_{kG} \to$ k-mod, which establishes an isomorphism

$$\mathbf{M}_k(G)^c \;\to\; \mathrm{Add}_k(\mathbf{H}_{kG}, \text{k-mod}).$$

The functor A is given by sending k[G/H] to $\mathbf{a}$(H) and a morphism f in k[H\G/K] to the morphism $f' \in \mathrm{Hom}_k(\mathbf{a}(H), \mathbf{a}(K))$, given by

$$HgL \;:\; \mathbf{a}(H) \to \mathbf{a}(K) \;:\; r \to r^g{}_{Hg \,\cap\, K}{}^K.$$

(9.44.) Let us give an example. Let D be a subgroup of G and put A(k[G/H]) = k[D\G/H], then multiplying (on the right) yields a functor A : $\mathbf{H}_{kG} \to$ k-mod (actually mod-k, but k-mod and mod-k are isomorphic, anyway). This functor corresponds to a G-functor (**a**, τ, ρ, σ) given by

$$\mathbf{a}(H) = k[D\backslash G/H],$$
$$\tau^K{}_H : k[D\backslash G/H] \to k[D\backslash G/K] : DgH \to DgK,$$
$$\rho_H{}^K : k[D\backslash G/K] \to k[D\backslash G/H] : DgK \to \Sigma_{K/H}DguH,$$
$$\sigma^f{}_H : k[D\backslash G/H] \to k[D\backslash G/H^f] : DgH \to DgfH^f.$$

Recall on the other hand that we have defined a G-functor for any V in kG-mod, by sending H $\in$ sub(G) to V^H. Let us call it c_V, then we obtain that the G-functor just defined is exactly $\mathbf{a} = c_{k[D\backslash G]}$.

(9.45.) As demonstrated in [Yo], the above isomorphism may be extended to arbitrary, not necessarily finite groups. In this case one has to restrict to subgroups of finite index which are pairwise commensurable, etc. As is already clear from the finite case, in order to get an additive functor T : $\mathbf{H}_G \to$ **(abelian groups)** = **Z**-mod, sending **Z**[G/H] to Pic(A^H, σ), it is necessary and sufficient to define a G-functor **a** over G (over **Z**), which sends H to Pic(A^H, σ), i.e. we have to define (compatible) restriction, corestriction and conjugation maps. For example, if H < L, then $A^L \subset A^H$ and we have to define maps Pic(A^L, σ) $\to$ Pic(A^H, σ) resp. Pic(A^H, σ) $\to$ Pic(A^L, σ). The first map is induced by $- \perp A^H$, the second should be a generalized norm map. Using this one then argues as in [RS] to derive (9.29.) for arbitrary groups.

(9.46.) To conclude, let us (very) briefly take a look at Brauer groups. We assume for simplicity's sake S to be a σ-Galois extension of R with Galois group G. We want to define a contravariant additive functor

$$T : \mathbf{H}_G = \mathbf{H}_{ZG} \to \textbf{(abelian groups)}$$

which sends $\mathbf{Z}[G/H]$ to $Br(S^H, s)$. One method to do this is to follow the lines of the proof of the analogous result for relative Picard groups, another uses the basic isomorphism $\mathbf{M}_k(G)^c = Hom_k(\mathbf{H}_{kG}, k\text{-mod})$.

So, let us make $\mathbf{a} : sub(G) \to \textbf{(abelian groups)} : H \to Br(S^H, \sigma)$ into a G-functor. It will be sufficient to define for every $H < K$ in $sub(G)$ and every $g \in G$ morphisms

$$\mathrm{cores}^H{}_K : Br(S^H, \sigma) \to Br(S^K, \sigma),$$
$$\mathrm{res}_K{}^H : Br(S^K, \sigma) \to Br(S^H, \sigma)$$
$$\mathrm{conj}^g{}_H : Br(S^H, \sigma) \to Br(S^{H(g)}, \sigma),$$

where $H(g) = H^g$ and to verify that these satisfy the necessary compatibilities, the Mackey formula and the cohomological axiom. Now, the definitions of the above morphisms are rather straightforward.

(i) $\mathrm{cores}_H{}^K = N^K{}_H$, the norm map $Br(S^H, \sigma) \to Br(S^K, \sigma)$ which is associated to $S^K \subset S^H$; in the σ-Galois case this reduces to $N^K{}_H(A) = (\perp_{K/H} A_g)^G$ where for any g representing a coset in K/H, we denote by A_g the g-twisted form of A;

(ii) $\mathrm{res}^K{}_H : Br(S^K, \sigma) \to Br(S^H, \sigma)$ is given by $[A] \to [A \perp S^H]$;

(iii) $\mathrm{conj}^g{}_H$ is induced by the isomorphism $g|S^H : S^H \to S^{H(g)}$, i.e. we put $\mathrm{conj}([A]) = [A \perp S^{H(g)}]$. We leave verifications to the reader, but refer to [For] for some indications on the proof (in the absolute case).

(9.47.) It is clear that the type of applications we gave for Picard groups and class groups is also valid for Brauer groups and divisorial Brauer groups. In particular, if $R \subset S$ is a divisorial Galois extension of Krull domains with Galois group G and if p is a prime which does not divide $|G/H|$ for some subgroup H of G, then the map

$$\beta(R)_{(p)} \rightarrow \beta(S^H)_{(p)}$$

is injective. If $\{H(i); 1 \leq i \leq n\}$ is a finite family of subgroups of G with the property that $\mathrm{g.c.d.}(\{|G/H(i)|; 1 \leq i \leq n\}) = 1$, then the map

$$\beta(R) \rightarrow \prod \beta(S^{H(i)})$$

is injective.

(9.48.) Let $\mathbf{H}(G, H) = \mathbf{Z}[H\backslash G/H] = \mathrm{End}_G(\mathbf{Z}[G/H])$ denote the <u>Hecke ring</u> associated to $H < G$, then the functors $T : \mathbf{H}_G \rightarrow$ **(abelian groups)** yield an action of $\mathbf{H}(G, H)$ on any $\mathrm{Pic}(S^H, \sigma)$ resp. $\mathrm{Br}(S^H, \sigma)$. In particular, these Hecke actions are compatible with restrictions $\sigma \leq \tau$, i.e. the group morphisms $\mathrm{Pic}(S^H, \sigma) \rightarrow \mathrm{Pic}(S^H, \tau)$ and $\mathrm{Br}(S^H, \sigma) \rightarrow \mathrm{Br}(S^H, \tau)$ are $\mathbf{H}(G, H)$ morphisms. It is thus easy to show that for any $\sigma \leq \tau$, the following is an exact sequence of $\mathbf{H}(G, H)$-modules:

$$\mathrm{Pic}(S^H, \sigma) \rightarrow \mathrm{Pic}(S^H, \tau) \rightarrow \mathrm{Bcl}(S^H; \sigma, \tau) \rightarrow \mathrm{Br}(S^H, \sigma) \rightarrow \mathrm{Br}(S^H, \tau).$$

In particular :

(9.49.) Proposition Let $R \subset S$ be a divisorially Galois extension of Krull domains with Galois group G, then for any $H < G$ there is an exact sequence of $\mathbf{H}(G, H)$-modules

$$0 \rightarrow \mathrm{Pic}(S^H) \rightarrow \mathrm{Cl}(S^H) \rightarrow \mathrm{Bcl}(S^H) \rightarrow \mathrm{Br}(S^H) \rightarrow \beta(S^H).$$

EXERCISES

The following is a short list of exercises related to the contents of the foregoing Sections. We have included only very few exercises on general sheaf theory or elementary algebraic geometry, since an enormous amount of material on these topics may be found in the many excellent textbooks on the subject, such as [Te, Ha, ...]. On the other hand, we have tried to include several open-ended exercises, which may be considered as" micro" research topics and by which we hope the reader will be stimulated to state and solve his own problems in the domains considered in this text.

1. Show by an example that for any E, F $\in$ $\underline{S}(X, \underline{O}_X)$ one does not necessarily have **Hom**(E, F)$_x$ = Hom(E$_x$, F$_x$).

2. Show that a subpresheaf of a sheaf need not be a sheaf.

3. Show that two sheaves on the same space can have the same stalks at every point, and still be different (even up to uisomorphism!).

4. Show by an example that for any pair of sheaves of $\underline{O}_X$-modules E, F, it is <u>not</u> necessarily true that sending U $\subset$ X to E(U) $\otimes$ F(U) yields a sheaf of $\underline{O}_X$-modules.

5. Let 0 $\to$ E' $\to$ E $\to$ E" $\to$ 0 be an exact sequence in $\underline{P}(X, \underline{O}_X)$. Prove the following assertions (cf. (2.35.)):

(i) if E' is a sheaf and E is separated, then E" is separated;

(ii) if E" is separated and E is a sheaf, then E' is a sheaf.

6. Let $(X, \underline{O}_X)$ be a ringed space. Give an example of an E $\in$ $\underline{S}(X, \underline{O}_X)$ such that there exists no surjective map of the form $(\underline{O}_X)^{(I)} \to$ E $\to$ 0 for any index set I.

For any U $\subset$ X, define $T_U\underline{O}_X$ as $(\underline{O}_X|U)^X$. Show that for any sheaf E in $\underline{S}(X, \underline{O}_X)$ we may find a family of open subsets U(i) together with an exact sequence

$$\oplus_i T_{U(i)}\underline{O}_X \to E \to 0.$$

7. Let $(X, \underline{O}_X)$ be a ringed space and $0 \to E' \to E \to E'' \to 0$ an exact sequence of $\underline{O}_X$-modules. Prove that if two sheaves in $\{E, E', E''\}$ are coherent, so is the third.

8. If E is locally free of finite rank and F is injective, then $E \otimes F$ is injective too. Prove this.

9. Let X be a prescheme and F, G sheaves of $\underline{O}_X$-modules. Show that if G is injective in $\underline{S}(X, \underline{O}_X)$, then **Hom**(F, G) is flabby.

10. (The projection formula) Let $f : (X, \underline{O}_X) \to (Y, \underline{O}_Y)$ be a morphism of ringed spaces. For any sheaf of $\underline{O}_X$-modules F and any locally free sheaf of $\underline{O}_Y$-modules E, prove that $f_*(F \otimes f^*E) \cong f_*F \otimes E$.

11. Let $f : X \to Y$ be a scheme morphism, then we call f dominating if f(X) is dense in Y. Prove that if f is dominating and if F is a torsionfree quasicoherent sheaf of $\underline{O}_X$-modules, then f_*F is a torsionfree sheaf of $\underline{O}_Y$-modules. Is this result valid if f is not dominating?

12. Let R be a noetherian ring and $X = \mathrm{Spec}(R)$. If T denotes the complement of the union of the minimal prime ideals of R, then the following assertions are equivalent:

(i) X(f) is a dense open subset of X;

(ii) f belongs to T.

Prove also that every dense open subset of X contains some X(f) of this form. Try to weaken the noetherian assumption.

13. Let $F'' \to F' \to E \to G' \to G''$ be an exact sequence of sheaves on a scheme X. Show that if F', F'', G', G'' are quasicoherent, then so is E.

14. Let $f : R \to S$ be a ring morphism and denote by $u : \mathrm{Spec}(S) \to \mathrm{Spec}(R)$ the induced map. Prove that u is injective if and only if the canonical map $\underline{O}_R \to u_*\underline{O}_S$ is injective.

15. A topological space X is noetherian if it satisfies the descending chain condition for closed subsets, i.e. if $Z_1 \supset Z_2 \supset \ldots$ is a chain of closed subsets of X, then there exists a positive integer n such that $Z_m = Z_n$ for all $m \geq n$.

Prove that the underlying topological space of any noetherian scheme is noetherian, but not conversely.

16. [GD] Let $(X, \underline{O}_X)$ be a locally ringed space and J a quasicoherent nilpotent sheaf of ideals of $\underline{O}_X$ such that

(i) $X_0 = (X, \underline{O}_X/J)$ is a scheme;

(ii) the sheaves of $\underline{O}_X/J$-modules J^k/J^{k+1} are quasicoherent.

Show that X is affine if and only if X_0 is affine.

17. Prove that any closed subscheme of an affine scheme is affine too.

18. [Ga] Let R be noetherian and σ an idempotent kernel functor in R-mod. Show $\mathbf{L}(\sigma) = \{I < R; \exists P_1, ..., P_n \in \mathbf{L}(\sigma) \cap \mathrm{Spec}(R), P_1. ... .P_n \subset I\}$. Does this remain valid if R is only assumed to be σ-noetherian?

19. [Gr2, Ste] Let R be a noetherian ring and M a finitely generated R-module, then define

$$\mathbf{L}(M, n) = \{I < R; \mathrm{Ext}^i(L, M) = 0 \text{ for all } i \leq n \text{ and all } L < R/I\}.$$

(i) Is $\mathbf{L}(M, n)$ a Gabriel filter? What if we assume R to be σ-noetherian and M just σ-finitely generated for some idempotent kernel functor σ in R-mod?

(ii) Show that the following assertions are equivalent:

(a) $I \in \mathbf{L}(M, 0)$;

(b) $[R/I, M] = 0$;

(c) I contains an M-regular element (i.e. some a with the property that $am = 0$ for some m in M, implies $m = 0$).

(iii) Show that the following assertions are equivalent:

(a) $I \in \mathbf{L}(M, n)$;

(b) $\mathrm{Ext}^i(R/I, M) = 0$ for all $i \leq n$.

What about property (c) in (ii)?

20. With notations as in the foregoing exercise, assume that R is a noetherian integrally closed domain and prove that the following assertions are equivalent:

(i) E is an **L**(R, 1)-torsion module;

(ii) $E_p = 0$ for every $p \in \mathrm{Spec}^{(1)}(R)$.

Is this also true for an arbitrary Krull domain.

21. Show that any injective object in the category (R, σ)-mod is also injective in R-mod.

22. Let R be a commutative ring, M an R-module and σ an idempotent kernel functor in R-mod. What is the relation between the modules σM and lim $[R/I, M]$, where I runs through **L**(σ)?

23. Let σ be an arbitrary idempotent kernel functor in R-mod and denote by $R \to R/\sigma R = R'$ the canonical projection. If σ' is the idempotent kernel functor in R'-mod induced by σ, then σ has property (T) if and only if σ' has property (T).

24. Let R be a σ-closed domain for some idempotent kernel functor s of finite type and M an R-module. Show that the following assertions are equivalent:

(i) $Q_\sigma(M)$ is torsionfree over R;

(ii) M_p is a torsionfree R_p-module for all $p \in \mathbf{K}(\sigma)$.

Can any of the hypotheses on R or σ be weakened? If not, give counterexamples.

25. [Ste] Let p be a prime ideal of R, which is assumed to be noetherian and let E be an injective hull of R/p. Let E_n consist of all e in E such that $p^n x = 0$. Prove that $E = \cup_n E_n$ and that E_{n+1}/E_n is a vector space over the field of fractions of R/p.

26. Show that if X is a separated quasicompact scheme and F a quasicoherent sheaf of $\underline{O}_X$-modules, which is locally of finite type, then there exists a sheaf of $\underline{O}_X$-modules G, which is locally of finite presentation and an surjection $u : G \to F$. Try to give a relative version of this.

27. Let X be a locally noetherian scheme and let B be a quasicoherent sheaf of $\underline{O}_X$-algebras, which is locally of finite type. Show that

(i) B is a coherent sheaf of rings;

(ii) a sheaf of B-modules F is coherent if and only if F is quasicoherent over $\underline{O}_X$ and locally of finite type over B.

Generalize this to locally Y-noetherian schemes.

28. [Bk, Car] Call a prime ideal p of R <u>soft</u> if the morphism $R \to R_p$ is surjective. Show that this property is equivalent to any one of the following:

(i) p is maximal and for any maximal ideal $q \neq p$, the ideals p and q are separated in Spec(R);

(ii) $R_p \otimes R_q = 0$ for all prime ideals $q \not\subset p$.

Let us call a sheaf of $\underline{O}_R$-modules E on X = Spec(R) <u>quasicoherent at</u> p if $E_p = \Gamma(X, E)_p$. Prove that the following assertions are equivalent :

(iii) p is soft;

(iv) all sheaves of $\underline{O}_R$-modules are quasicoherent at p;

(v) the sheaf G defined by $\Gamma(U, G) = \Pi_{q \in U}(R_p \otimes R_q)$ is quasicoherent at p.

29. Let X be a noetherian scheme and let K(X) denote the Grothendieck group of X, i.e. the quotient of the free abelian group on isomorphism classes of coherent sheaves on X over the subgroup generated by the [E] - [E'] - [E''], where

$$0 \to E' \to E \to E'' \to 0$$

is an exact sequence of coherent sheaves. Show that if Z is close, with complement Y = X - Z, then there is an exact sequence

$$K(Z) \to K(X) \to K(Y) \to 0.$$

Deduce a similar result if Y is just a generically closed subset of X (in this case, you will have to work with relative K-groups $K_Y(X)$, where one uses Y-coherent sheaves and exact sequences in the quotient category $\underline{S}_Z(X, \underline{O}_X)$, etc.).

30. [Ha, ...] Let I be an ideal of a noetherian ring R and E an injective R-module. Prove that $\sigma_I E$ is an injective R-module as well. (Hint : Use Krull's Theorem!) What happens if R is not necessarily noetherian?

31. If E is injective in $\underline{S}(X, \underline{O}_X)$ resp $\mathbf{Q}(X, \underline{O}_X)$, then for any open subset U of X the restriction E|U is injective in $\underline{S}(U, \underline{O}_X|U)$ resp. $\mathbf{Q}(U, \underline{O}_X|U)$. What about E|Y, if Y is a generically closed subset of a (locally noetherian) scheme X?

32. Prove that $U = X - \{(0, 0)\} \subset X = \mathbf{A}^2(\mathbf{C})$ is not affine.

33. [Ha, ...] Let R be a noetherian ring, E an injective R-module and f an element of R. Prove that the localization morphism $E \to E_f$ if surjective. As a corollary, show that the sheaf $\underline{O}_E$ on Spec(R) is flabby.

34. Let R be σ-noetherian and E injective in (R, σ)-mod. Prove that E_p is injective in R_p-mod for all $p \in \mathbf{K}(\sigma)$.

35. Prove the following stronger version of (6.40.) : assume X to be locally Y-noetherian for some generically closed subset Y, then $Cl_{X/Z}(E) = Q_Y(E)$ for any quasicoherent sheaf of $\underline{O}_X$-modules E on X.

36. Let $(X, \underline{O}_X)$ be a locally ringed space and F, F' sheaves of $\underline{O}_X$-modules, which are both locally of finite type. Prove that:

(i) $Supp(F) = \{x \in X; F_x \neq 0\}$ is closed in X;

(ii) $Supp(F \otimes F') = Supp(F) \cap Supp(F')$.

37. Let X be a scheme, J a sheaf of $\underline{O}_X$-ideals and F a quasicoherent sheaf, which is locally of finite type on X. Prove that $Supp(F/JF) = Supp(F) \cap V(J)$.

38. Let $(X, \underline{O}_X)$ be a Krull scheme and consider an exact sequence of sheaves of $\underline{O}_X$-modules

$0 \to E' \to E \to E'' \to 0.$

Show that if E' is normal and E is reflexive, then E" is torsionfree. How may the assumptions on X be weakened?

39. Let X be a Krull scheme and let E be a reflexive sheaf on X. Assume $F \subset E$ is a lattice on X, then a lattice F' with $F \subset F' \subset E$ and rank(F) = rank(F') is said to be an <u>extension</u> of F in E. If F' is also normal, then we speak of a <u>normal extension</u> of F in E.

If Q = E/F and T denotes the torsion subsheaf of Q, then we let $F^!$ denote the kernel of the obvious map $E \to Q/T$. Prove that:

(i) $F^!$ is the largest extension of F in E;

(ii) $F^!$ is a normal extension of F in E.

(This strengthens a similar result in [OSS].)

40. Let Y be a generically closed subset of a scheme X which is locally noetherian with respect to some generically closed subset Y and let E be a quasicoherent resp. coherent sheaf on X. What can you say about $H^0_Y E$? What if we drop the noetherian hypothesis? What about $H^i_Y E$?

41. [Ha2] If $(X, \underline{O}_X)$ is locally factorial, then every reflexive rank 1 sheaf of $\underline{O}_X$-modules is invertible. Prove this. (Note : this is stronger than just asserting that Cl(X) = Pic(X)!)

42. [Ru] Let R be a domain with field of fractions K and assume that R $= \cap R_p$, where p runs through some subset $Y \subset \mathrm{Spec}^{(n)}(R)$, then for any ideal I with the property that ht(I) > n, prove that $I^{-1} = R$, where as usually, $I^{-1} = \{a \in K; aI \subset R\}$.

43. [Ru] If R is a Krull domain and every idempotent kernel functor in R-mod has property (T), then R is a Dedekind domain.

44. [Tr] Let $X = \text{Spec}(R)$ be a 1-regular affine scheme, then $S = \cap_{ht(p)=1} R_p$ is a Krull domain and for $Y = \text{Spec}(S)$ we have $(X^{(1)}, \underline{O}_X|X^{(1)}) \cong (Y^{(1)}, \underline{O}_Y|Y^{(1)})$. Prove this.

45. [LO1] Let X be a Krull scheme and let E, F be torsionfree sheaves of $\underline{O}_X$-modules, then $E \otimes F \cong E \perp F$ whenever E is flat (i.e. E_x is a flat $\underline{O}_{X,x}$-module for any $x \in X$).

46. Prove that any integral or flat extension of Krull domains $B \to A$ satisfies PDE.

47. Let R be a Krull domain with field of fractions K and M an R-lattice. Show that the following assertions are equivalent:

(i) M is divisorial;

(ii) $\text{Ann}(Rx)$ is a divisorial ideal for all $x \in KM/M$.

Relative version?

48. Let Z be an irreducible closed subset of a Krull scheme X, determined by the sheaf of ideals J of $\underline{O}_X$ and consider the obvious map $f : \text{Pic}(X) \to \text{Pic}(X, Y)$, where Y is the complement of Z in X. It has been proved in [VV2] that the kernel of this map is generated by the isomorphism classes of invertible sheaves amongst the $(J^n)^{**}$, in the noetherian affine case. Is this also true in the general situation?

49. Prove that $\Phi : \mathbf{Z}[H\backslash G/K] \to \text{Hom}_G(\mathbf{Z}[G/H], \mathbf{Z}[G/K])$ which sends HxK to the map given by $gH \to \Sigma\, guK$, where u runs through a set of representatives of HxK/K defines a bijective correspondence between these sets.

50. Let A be a R-algebra and G a group acting as R-automorphisms on A. Show by an example that if G is not finite, then we do not necessarily have that $(A^G)_f = (A_f)^G$ for any f in R.

51 - 118. Solve the exercises in [VV2] (if you didn't already).

REFERENCES

[AB1] Albu T., Nastasescu C., Décompositions Primaires dans les Catégories de Grothendieck Commutatives 1, J. Reine Angew. Math. 280 (1976) 172-194.

[AB2] Albu T., Nastasescu C., Décompositions Primaires dans les Catégories de Grothendieck Commutatives 2, J. Reine Angew. Math. 282 (1976) 172-185.

[An] Anderson F.W., Endomorphism Ringes of Projective Modules, Math. Z. 111 (1969) 322-332.

[AF] Anderson F.W., Fuller K., Rings and Categories of Modules, Springer Verlag, Berlin, 1974.

[AM] Atiyah M., MacDonald I.G., Introduction to Commutative Algebra, Addison-Wesley, Reading, Mass., 1969.

[Au1] Auslander B., The Brauer Group of a Ringed Space, J. of Algebra 4 (1966) 220-273.

[Au 2] Auslander B., Finitely generated reflexive modules over integrally closed noetherian domains, ph.d. thesis, Univ. of Michigan, 1963.

[Au3] Auslander B., Central separable algebras which are locally endomorphism rings of free modules, Proc. Amer.Math.Soc. 30 (1971) 395-404.

[AG1] Auslander M., Goldman O., Maximal Orders, Trans. Amer. Math. Soc, 97 (1960) 1-24

[AG2] Auslander M., Goldman O., The Brauer Group of a Commutative Ring, Trans. Amer. Math. Soc, 97 (1960) 376-409.

[Az] Azumaya G., On maximally central algebras, Nagoya Math. J. 2 (1951) 119-150.

[Bar] Barth W., Some properties of stable rank 2 vector bundles on $\mathbf{P}^n$, Math. Ann. 226 (1977) 125-150.

[Ba] Bass H., Algebraic K-Theory,Benjamin, New York 1968.

[Ba1] Bass H., Lectures on Topics in Algebraic K-theory, Tata Institute of Fundamental Research, Bombay, 1967.

[Ba2] Bass H., Torsion free and projective modules, Trans. Amer. Math. Soc. 102 (1962) 319-327.

[BM] Bass H., Murthy P., Grothendieck Groups and Picard Groups of Abelian Group Rings, Ann. of Math. 86 (1967) 16-73.

[Be] Beck I., Injective Modules over a Krull Domain, J. Algebra 17 (1971) 116-131.

[Bz1] Bijan-Zadeh M.H., Torsion Theories and Local Cohomology over Commutative Noetherian rings, J. London Math. Soc. 19 (1979) 402-410.

[Bz2] Bijan-Zadeh M.H., A common generalization of local cohomology theories, Glasgow Math. J. 21 (1980) 173-181.

[Bk] Bkouche R., Thèse, Bull. Soc. Math. France 98 (1970) 253-295.

[Bo] Bourbaki N., Algèbre Commutative, Eléments de Math. 27, 28, 30, 31, Hermann, Paris, 1961-66.

[Br] Bredon G., Sheaf Theory, MacGraw-Hill, New York, 1967.

[CV] Caenepeel S., Verschoren A., A relative version of the Chase-Harrison-Rosenberg exact sequence, J. Pure Appl. Algebra (to appear).

[Ca] Cahen P-J., Commutative Torsion Theory, 184 (1973) 73-85.

[Car] Carral M., K. Theory of Gelfand Rings J. Pure Appl. Algebra 17 (1980) 249-265.

[CE] Cartan H., Eilenberg S., Homological Algebra, Princeton University Press, Princeton 1956.

[CHR] Chase S., Harrison P., Rosenberg A., Galois Theory and Cohomology of Commutative Rings, Mem. Amer. Math. Soc. 52 (1965) 1-19.

[CGO] Childs L., Garfinkel G., Orzech M., On the Brauer Group and Factoriality of Normal Domains, J. Pure Appl. Algebra 6 (1975) 111-123.

[CF] Claborn L., Fossum R., Class Groups of n-noetherian Rings, J. of Algebra 8 (1968) 265-285.

[CF1] Claborn L., Fossum R., Generalizations of the notion of class group, Ill. J. Math. 12 (1968) 228-253.

[DI] De Meyer F., Ingraham E., Separable Algebras over Commutative Rings, LNM 181, Springer Verlag, Berlin, 1971.

[For] Ford T., Hecke actions on Brauer groups and étale cohomology groups, preprint.

[Fos] Fossum R.M., The Divisor Class Group of a Krull Domain, Springer Verlag, Berlin 1973.

[Fos1] Fossum R., Maximal orders over Krull domains, J. Algebra 10 (1986) 321-332.

[Ga] Gabriel P., Des Catégories Abeliennes, Bull. Soc. Math. France, 90 (1962) 323-448.

[GP] Gabriel P., Popescu N., Caractérisation des catégories abeliennes avec générateurs et limites inductives exactes, C.R.Acad.Sc. Paris 258 (1964) 4188-4190.

[Go] Godement R., Topologie Algébrique et Théorie des Faisceaux, Hermann, Paris, 1958.

[Gola] Golan J., Localization in Noncommutative Rings, M. Dekker, New York, 1975.

[Gold] Goldman O., Rings and Modules of Quotients, J. Algebra 13 (1969) 10-47.

[Gri] Griffiths P.A., The Brauer group of A[T], Math. Z. 147 (1976) 79-86.

[Gr1] Grothendieck, Sur Quelques Points d'Algèbre Homologique, Tohoku Math. J. 9 (1957) 119-221.

[Gr2] Grothendieck A., Le Groupe de Brauer, I. Algèbres d'Azumaya et Interprétations Diverses. II. Théorie Cohomologique. III. Exemples et Compléments. in: Dix Exposés sur la Cohomologie des Schémas, North Holland, Amsterdam, 1968 pp. 46-188.

[Gr3] Grothendieck A., Local Cohomology, LNM 41, Springer Verlag, Berlin, 1967.

[GD] Grothendieck A., Dieudonné J., Eléments de Géométrie Algébrique 1, Springer Verlag, Berlin, 1971.

[Ha] Hartshorne R., Algebraic Geometry, Springer Verlag, Berlin, 1977.

[Ha1] Hartshorne R., Residues and Duality, LNM 20, Springer Verlag, Berlin, 1966.

[Ha2] Hartshorne R., Stable reflexive sheaves, Math. Ann. 254 (1980) 121-176.

[Ho] Horrocks G., Vector bundles on the punctured spectrum of a local ring, Proc. London Math. Soc. 14 (1964) 689-713.

[KO1] Knus M.A., Ojanguren M., Théorie de la Descente et Algèbres d'Azumaya, LNM 389, Springer Verlag, Berlin, 1974.

[KO2] Knus M.A., Ojanguren M., A Norm for Modules and Algebras, Math. Z. 142 (1975) 33-45.

[La] Lazard D., Autour de la Platitude, Bull. Soc. Math. France 97 (1969) 81-128.

[LO1] Lee H., Orzech M., Brauer Groups, Class Groups and Maximal Orders for a Krull Scheme, Can. J. Math. 34 (1982) 996 - 1010.

[LO2] Lee H., Orzech M., Brauer groups and Galois cohomology for a Krull scheme, J. Algebra 95 (1985) 309-331.

[Ma] MacDonald I.G., Algebraic Geometry, Introduction to Schemes, W.A. Benjamin, New York, 1968.

[Mat] Matsumura H., Commutative Algebra, Benjamin, New York, 1970.

[Mil] Milne J.S., Etale Cohomology, Princeton University Press, Princeton, New Jersey, 1980.

[Mi] Mitchell B., Theory of Categories, Academic Press, New York, 1965.

[MVO] Murdoch D., Van Oystaeyen F., Noncommutative Localization and Sheaves, J. Algebra 35 (1975) 500-515.

[Na] Nastasescu C., La Structure des modules par rapport à une topologie additive, Tohoku Math. J. 26 (1974) 173-201.

[No] Northcott D.G., An introduction to homological algebra, Cambridge University Press, Cambridge, 1960.

[Or1] Orzech M., Brauer Groups and Class Groups for a Krull Domain, in: "Brauer Groups in Ring Theory and Algebraic Geometry", LNM 917, Springer Verlag, Berlin, 1982.

[Or2] Orzech M., Divisorial Modules and Krull Morphisms, preprint 1981-82, Queen's University, 1981.

[OS] Orzech M., Small C., The Brauer Group of a Commutative Ring, M. Dekker, New York, 1975.

[OSS] Okonek C., Schneider M., Spindler H., Vector bundles on complex projective spaces, Birkhäuser Verlag, Basel, Boston, Stuttgart, 1980.

[OV] Orzech M., Verschoren A., Some Remarks on Brauer Groups of Krull Domains, LNM 917, Springer Verlag, Berlin, 1982, pp. 91-96.

[Pa] Pareigis B., Non-additive ring and module theory IV : The Brauer Group of a Symmetric Monoidal Category, in "Brauer Groups", Evanston 1975, LNM 549, Springer Verlag, Berlin, 1975.

[Pe] Perlis R., On the class numbers of arithmetically equivalent fields, J. Number Theory 10 (1978) 489-509.

[Re] Reiner I., Maximal Orders, Academic Press, New York, 1975.

[Ri] Riley J., Reflexive ideals in maximal orders, J. Algebra 2 (1965) 451-465.

[RS] Roggenkamp K., Scott L., Hecke Actions on Picard Groups, J. Pure Appl. Algebra 26 (1982) 85-100.

[Ru] Rubin, R. On exact localization, Pac. J. Math, 49 (1973) 473-481.

[Se] Serre J-P., Faisceaux Algébriques Cohérents, Annals of Math. 61 (1955) 197-278.

[So] Sonn J., Class groups and Brauer groups, Israel J. Math. 34 (1979) 97-106.

[Ste] Stenstrøm B., Rings of Quotients, Springer Verlag, Berlin 1975.

[Str] Strooker J., Introduction to Categories, Homological Algebra and Sheaf Cohomology, Cambridge University Press, Cambridge, 1978.

[Su] Suominen, K., Localization of sheaves and Cousin complexes, Acta Mathematica 131 (1973) 27-41.

[Sw1] Swan R., The Theory of Sheaves, Chicago University Press, Chicago, 1964.

[Sw2] Swan R., Algebraic K-Theory, LNM 76, Springer Verlag, Berlin, 1968.

[Te] Tennison B.R., Sheaf Theory, London Math. Soc. Lect. Notes Series 20, Cambridge University Press, Cambridge, 1976.

[Tr] Treger R., Reflexive Modules, J. Algebra 54 (1978) 444-466.

[VO1] Van Oystaeyen F., Compatibility of kernel functors and localization functors, Bull. Soc. Math. Belg. 18 (1976) 131 - 137.

[VO2] Van Oystaeyen F., Prime Spectra in Non-commutative Algebra, LNM 444, Springer Verlag, Berlin, 1975.

[VV1] Van Oystaeyen F., Verschoren A., Reflectors and Localization, M. Dekker, New York, 1979.

[VV2] Van Oystaeyen F., Verschoren A., Relative Invariants of Rings. The commutative theory, M. Dekker, New York, 1983.

[VV3] Van Oystaeyen F., Verschoren A., Relative Invariants of Rings. The noncommutative theory, M. Dekker, New York, 1984.

[VV4] Van Oystaeyen F., Verschoren A., Extending Coherent and Quasi-Coherent Sheaves on Generically Closed Spaces, J. of Algebra 89 (1984) 224 - 236.

[Ver] Verdier, J-L., Topologies et faisceaux, Sém. Artin-Grothendieck-Verdier 1963-64, LNM 269, Springer Verlag, Berlin 1972.

[Ve1] Verschoren A., On the Picard Group of a Grothendieck Category, Comm. Algebra 8 (1980) 1169-1194.

[Ve2] Verschoren A., The Brauer Group of a Quasi-affine Scheme, in: "Brauer Groups in Ring Theory and Algebraic Geometry", LNM 917, Springer Verlag, Berlin, 1982.

[Ve3] Verschoren A., On the Reflexive Class Group, Indag . Math 45 (1983) 111-115.

[Ve4] Verschoren A., Exact Sequences for Relative Brauer Groups and Picard Groups, J. Pure Appl. Algebra 28 (1983) 93-108.

[Ve5] Verschoren A., Norms and Generalized Class Groups, J. Pure Appl. Algebra (submitted).

[Ve6] Verschoren A., Global Relative Invariants, to appear.

[Ve7] Verschoren A., Brauer groups and class groups of Krull domains : a K-theoretic approach, J. Pure Appl. Algebra 33 (1984) 219-224.

[Ve8] Verschoren A., Tertiary decomposition in Grothendieck categories, Czech. Math. J. 30 (1980) 661-672.

[Yo] Yoshida T., On G-Functors (II): Hecke operators and G-functors, J.Math.Soc.Japan 35 (1983) 179-190.

[Yu] Yuan S., Reflexive modules and algebra class groups over noetherian integrally closed domains, J. Algebra 32 (1974) 405 - 417.

[ZS] Zariski O., Samuel P., Commutative Algebra, 2 vols., van Nostrand, Princeton, 1958/1960.

[Ze] Zelinsky D., Long Exact Sequences and the Brauer Group, Evanston 1975, LNM 549, Springer Verlag, Berlin, 1976, pp. 63-70.

INDEX